BIG RED STATISTICS

Written by Mike Davidson

Edited by Stephen Shone

Name: Luke Boyer
luke.boyer20@montgomerybell.edu
Shone
615-788-5488

Big Red Statistics

Copyright © 2017 by Michael Davidson

All rights reserved.

This book or any portion thereof may not be reproduced or used in any manner whatsoever without the express written permission of the publisher except for the use of brief quotations in a book review or scholarly journal.

First Printing: 2016

Big Red Press
4001 Harding Road

Nashville, TN 37205

www.montgomerybell.edu

Special thanks to Jamie Tillman, Eibhlin Colgan, Brad Gioia and the Davis Family, who helped make this project a reality.

Contents

Introduction 1
    Chapter 1. Statistics: Working with Data 1
        Class Activity 1.1: Vanderbilt Men's Basketball Team 4
        Class Activity 1.2: Sample Surveys 5
        Homework Assignment 1.1: Junior School Sample Survey 5
        Class Activity 1.3: Comparing Survey Results 7
        Class Activity 1.4: Experimentation 7
        Homework Assignment 1.2: Chapter 1 Review 9

Part A: Producing Data 11
    Chapter 2. Sampling for Observational Studies 11
        Class Activity 2.1: A Fireworks Display for MBA's Sesquicentennial? 11
        Homework Assignment 2.1: Bias and Simple Random Samples 15
        Class Activity 2.2: Random Numbers and Pseudo-Random Numbers 16
        Class Activity 2.3: Sample Size and Variability 16
        Homework Assignment 2.2: Variability with a Larger Sample Size 19
        Class Activity 2.4: A Quick Method for Margin of Error 21
        Homework Assignment 2.3: Working With Margin of Error 25
        Class Activity 2.5: Population Size vs. Sample Size 25
        Class Activity 2.6: Alternatives to Simple Random Samples 26
        Homework Assignment 2.4: Stratified Random Samples and Cluster Samples 27
        Class Activity 2.7: Sampling Errors 28
        Homework Assignment 2.5: Sampling Errors in Review 29
    Chapter 3. Experimenting 31
        Class Activity 3.1: A Headache Outbreak at MBA 31
        Homework Assignment 3.1: Redesigning the Herbal Tea Experiment 33
        Class Activity 3.2: Other Issues with Clinical Trials 34
        Class Activity 3.3: Statistical vs. Practical Significance 35
        Homework Assignment 3.2: Types of Significance 36
        Class Activity 3.4: Other Experimental Designs 37
        Homework Assignment 3.3: Testing the Effect of an MBA Education 40

Class Activity 3.5: Data Ethics 41

Class Activity 3.6: The Tuskegee Study 41

Class Activity 3.7: The Milgram Experiment 42

Class Activity 3.8: The Stanford Prison Experiment 43

Homework Assignment 3.4: An Ethical Question 44

Part B: Analyzing Data 45

Chapter 4. Displaying and Describing Data 45

Class Activity 4.1: Dotplots 48

Class Activity 4.2: Stemplots 49

Class Activity 4.3: Histograms 50

Homework Assignment 4.1: Displaying Distributions Graphically and Interpreting Histograms 51

Class Activity 4.4: Making a Histogram on Your Calculator 54

Class Activity 4.5: Beginning to Describe Distributions 55

Homework Assignment 4.2: Describing and Graphing Batting Averages 57

Class Activity 4.6: Describing Distributions with Numbers 58

Homework Assignment 4.3: Making an MBA Football Boxplot 61

Class Activity 4.7: Making a Boxplot on Your Calculator 62

Class Activity 4.8: Mean and Standard Deviation 63

Homework Assignment 4.4: Finding Standard Deviation with Excel 65

Class Activity 4.9: Finding s on Your Calculator 70

Class Activity 4.10: Standard Deviation for Samples and Populations 70

Class Activity 4.11: Misrepresentation of Data 71

Homework Assignment 4.5: Lies, Damned Lies and Line Graphs 74

Chapter 5. Studying Individuals within a Distribution 75

Class Activity 5.1: Finding Percentile 75

Class Activity 5.2: Using Standardized Tests to Understand Z-Scores 76

Class Activity 5.3: Collecting Data about Theater Attendance 76

Homework Assignment 5.1: Interpreting Z-Scores and Percentiles 77

Class Activity 5.4: Exploring Density Curves 78

Homework Assignment 5.2: Sketching Density Curves and Approximating Mean and Median . 80

Class Activity 5.5: Normal and Uniform Distributions 82

Class Activity 5.6: Distributions That Are Normal 83

Class Activity 5.7: Is it a Normal Distribution? Are You Sure? 84

Homework Assignment 5.3: Is it a Uniform Distribution? Are you Sure? 86

Class Activity 5.8: First Digits and Final Digits 87

Class Activity 5.9: The 68-95-99.7 Rule 88

Homework Assignment 5.4: Examining the "New SAT" 90

Class Activity 5.10: The Standard Normal Distribution 91

Class Activity 5.11: Brandt Snedeker at the Driving Range 93

Homework Assignment 5.5: Some Proportion Problems 95

Class Activity 5.12: The Prom Date Problem 96

Chapter 6. Bivariate Data Analysis 101

Class Activity 6.1: Our First Scatterplot: Height vs. Wingspan 101

Class Activity 6.2: Distance to School and Commute Time 103

Homework Assignment 6.1: Questions about Your Morning Drive 104

Class Activity 6.3: Finding Correlation 105

Class Activity 6.4: Finding Correlation on Your Calculator 106

Homework Assignment 6.2: Exploring Correlation (r) 108

Class Activity 6.5: Understanding Least-Squares Regression 110

Class Activity 6.6: Doing Least-Squares Regression on a Calculator 111

Homework Assignment 6.3: Exploring Regression and Prediction 114

Class Activity 6.7: Understanding the Coefficient of Determination (r^2) 115

Class Activity 6.8: Residuals and Residual Plots 117

Homework Assignment 6.4: The Coefficient of Determination vs. Residual Plots 120

Class Activity 6.9: Correlation vs. Causation 121

Homework Assignment 6.5: Exploring Causation 122

Class Activity 6.10: Bivariate Data Project 124

Part C: Probability 127

Chapter 7. Probability Rules 127

Class Activity 7.1: Randomness and Probability 127

Homework Assignment 7.1: Exploring The Law of Large Numbers 129

Class Activity 7.2: Simulation 130

Homework Assignment 7.2: Running a Simulation on Excel 131

Class Activity 7.3: Probability Models 133

Class Activity 7.4: Two-Way Tables 134

Homework Assignment 7.3: Using a Two-Way Table 134

Class Activity 7.5: Venn Diagrams 135

Homework Assignment 7.4: Making and Interpreting a Venn Diagram 136

Class Activity 7.6: Conditional Probability and Tree Diagrams 137

Homework Assignment 7.5: Interpreting Tree Diagrams 139

Class Activity 7.7: Determining Our General Probability Rules 140

Class Activity 7.8: Using Our General Probability Rules 141

Homework Assignment 7.6: Introducing Simpson's Paradox 142

Class Activity 7.9: Drug Testing at MBA and Conditional Probability 143

Class Activity 7.10: 5 Seconds on the Clock.. 144

Class Activity 7.11: Does the "Hot Hand" Exist? 145

Homework Assignment 7.7: Do You Have the Hot Hand? 146

Chapter 8. Modeling with Probability 147

Class Activity 8.1: Expected Value... 147

Class Activity 8.2: Expected Value with Tree Diagrams 148

Homework Assignment 8.1: When Should You Buy Insurance? 150

Class Activity 8.3: Random Variables and a Brown's Diner Burger 151

Homework Assignment 8.2: Octuplets Walk Into Brown's Diner 153

Class Activity 8.4: Brown's Catfish Dinner and Counting 101 154

Homework Assignment 8.3: Multiplication Principle and Factorials 155

Class Activity 8.5: Permutations and Combinations in Brentwood 156

Homework Assignment 8.4: Mr. D's Ice Cream Shoppe and Poker Probabilities Revisited 158

Class Activity 8.6: The Binomial Theorem at the UT-Vandy Game 159

Homework Assignment 8.5: Is it a Binomial Setting? 162

Class Activity 8.7: Solving the Catfish Problem 163

Class Activity 8.8: Binomial and Sampling Distributions 164

Homework Assignment 8.6: Working With Sampling Distributions 166

Class Activity 8.9: The Boy Named Banjo Concert 167

Homework Assignment 8.7: What about the Martingale? 170

Class Activity 8.10: Kelly Criteria: What to Do When You Have the Edge 171

Homework Assignment 8.8: Chapter 8 Review 172

Part D: Inference 175

Chapter 9. Confidence Intervals for Proportions 175

Class Activity 9.1: A More Perfect Confidence Interval 175

Homework Assignment 9.1: Where Did 1/√n Come From? 176

Class Activity 9.2: What Happens When p ≠ .5? 177

Class Activity 9.3: What If We Don't Know p? 178

Homework Assignment 9.2: A More Accurate 95% Confidence Interval 179

Class Activity 9.4: Different Levels of Confidence 180

Class Activity 9.5: Finding Levels of Confidence 181

Homework Assignment 9.3: Captain Norton and Service 183

Chapter 10. Significance Tests and Confidence Intervals for Means 185

Class Activity 10.1: Free Throws at the Faculty-Student Game 185

Class Activity 10.2: Tap Water vs. Lime Spa Water: One-Sided and Two-Sided Alternative Hypotheses 187

Homework Assignment 10.1: Finding P-values on Your Calculator 188

Class Activity 10.3: Testing Our Own Free Throw Claims 191

Homework Assignment 10.2: What Can We Conclude About Our Free Throw Shooting? 191

Class Activity 10.4: Is Mr. Davidson's Die Fair? 195

Homework Assignment 10.3: Making a Chi-Square Distribution for Five Degrees of Freedom 197

Class Activity 10.5: Using the Chi-Square Distribution to Perform a Chi-Square Test for Goodness of Fit 198

Homework Assignment 10.4: Testing the Tootsie Pop Distribution 201

Class Activity 10.6: The Chi-Square Test for Homogeneity and Independence 203

Homework Assignment 10.5: Using a Calculator to Perform Two-Way Table Chi-Square Test 205

Class Activity 10.7: Inference about a Population Mean: Ages at a Little League Game 205

Homework Assignment 10.6: The Central Limit Theorem and Standard Deviation of a Sampling Distribution 207

Class Activity 10.8: Making a Confidence Interval for a Population Mean 207

viii

Homework Assignment 10.7: Practicing Mean Confidence Intervals 209

Class Activity 10.9: The T Interval: Better for Small Samples and Larger 210

Homework Assignment 10.8: A Better Confidence Interval With t* 211

Class Activity 10.10: The T Distribution on the Calculator 211

Homework Assignment 10.9: Chapter 10 Review Exercises 212

Index 215

Tables 218

Introduction

Chapter 1. Statistics: Working with Data

Who will be the next Governor of Tennessee? How about the next senior class president at MBA? In the first quarter of a football game, should MBA go for a two-point conversion after a touchdown against Father Ryan? Do straight plays or musicals draw bigger crowds for the MBA theatre department? Is a new drug effective in lowering the risk of a heart attack? Do mouthguards reduce the risk of concussions in soccer games?

People often seek to answer questions like these using anecdotal evidence. In the game against Baylor the two-point conversion was successful, so a fan thinks it will work against Father Ryan as well. A student recalls sold out crowds at the two musicals he attended last year, so he is convinced that musicals draw bigger crowds. In both of these cases, the conclusion is based on data but the sample size is small and its collection is haphazard.

Scientists answer questions like these by collecting data through **observational studies** and **experiments**. Observational studies are those in which subjects' responses are not intentionally influenced. Experiments are studies where a treatment is intentionally imposed for the purpose of observing the response to that treatment. If an APES class went into the forest to listen to bird calls, the presence of the class might change the behavior of the birds but since any change created is not intentional, it would still qualify as an observational study. If the APES class brought a stereo to see how birds would respond to different types of music, then they would be performing an experiment. **Statistics** is the study of how best to collect, analyze, interpret and present data.

Tim, a junior at MBA, is looking at colleges. He is only interested in 4-year colleges in the United States. There are, however, over 2,500 4-year colleges in the United States so to start his search Tim heads to the college counseling office to see which colleges have received the most applications from MBA recently.[1] There are eighteen schools that had a dozen or more applications from the class of 2015. Tim would like to be close to home, but not too close, he would prefer a school with Division I athletics and since he is not sure what he wants to pursue in college, would prefer a school with a large number of majors offered. Tim does some research and makes the following table.[2]

College	Miles from Nashville (car)	Athletic Division	Number of Majors
University of Alabama	245	I	71
Auburn University	301	I	144
Furman University	341	I	43
University of Georgia	291	I	186
University of Mississippi	233	I	69
University of North Carolina	512	I	59
University of Pennsylvania	800	I	110
Princeton University	853	I	31
Rhodes College	205	III	34
Sewanee	92	III	37
University of South Carolina	443	I	66
Southern Methodist University	667	I	66
University of Tennessee-Knoxville	179	I	124
Texas Christian University	699	I	89
Vanderbilt University	0	I	67
University of Virginia	544	I	48
Wake Forest	436	I	51
Washington University in St. Louis	316	III	144

Table 1.1

In this table, the **individuals** are 4-year colleges. Individuals are the objects being described by a set of data. The three **variables** that most concerned Tim were distance to the school from Nashville, the athletic division the school competed in, and the number of majors offered. A **variable** is a characteristic of an individual. There are **categorical variables** (also called qualitative variables) and **quantitative variables**. A categorical variable is one that fits into one of a number of possible groups. A quantitative variable is one that takes a numerical value. Quantitative variables are further divided into **discrete variables** and **continuous variables**. A discrete variable can only take on certain values; a continuous variable can take on any value within a range.

In this table, athletic division is a categorical variable. There are four possible groups or categories into which a school can be placed: Division I, Division II, Division III or no NCAA athletics. It might be tempting to think that athletic divisions are quantitative because of the numbers 1, 2, and 3. If a variable is quantitative, it should be coherent to compare one variable's value to another. It does not make sense to say that Sewanee's athletic division is three times that of Vanderbilt's.

It does, however, make sense to say that when driving from Nashville, Auburn University is 56 miles farther away than The University of Alabama and to say that The University of Pennsylvania has more than three times as many majors offered as Princeton University. This is because miles from Nashville by car and number of majors offered are both quantitative variables.

Miles from Nashville is a continuous variable. Even though the table rounded the distance to the nearest mile, it is possible for the distance from Nashville to be any value greater than or equal to 0 miles. A college can be 138.2 miles away, or 138.211 miles if you want to be very precise in your measurement. Number of majors, however, is a discrete variable because only whole number values are possible. It is possible to offer 66 majors, like the University of South Carolina, or 67 majors, like Vanderbilt University, but it would be impossible for a college to offer 66.4 majors.

Class Activity 1.1: Vanderbilt Men's Basketball Team

The following is the 2014-15 Vanderbilt Men's Basketball Roster:[3]

Name	Class	Position	Height	Weight (lbs)
Wade Baldwin IV	FR	G	6'3"	195
Nolan Cressler	RS SO	G	6'4"	204
Matthew Fisher-Davis	FR	G/F	6'5"	173
Josh Henderson	RS SR	C	6'11"	240
Damian Jones	SO	C	6'10"	248
Carter Josephs	JR	G	6'0"	180
Luke Kornet	SO	F	7'0"	240
Riley LaChance	FR	G	6'2"	194
Phillip McGloin	FR	G	6'1"	178
Shelton Mitchell	FR	G	6'3"	186
Shelby Moats	SR	F	6'8"	221
Jeff Roberson	FR	F	6'6"	206
James Siakam	SR	F	6'7"	225
Nathan Watkins	JR	G	6'5"	195

What individuals does this data set describe? *the Vandy players*

What variables given are categorical? *Class & Position*

What variables given are quantitative? *Height & Weight*

What is a categorical variable that is not shown? *dominant shooting hand*

...a discrete quantitative variable that is not shown? *Total # of shots that season*

...a continuous quantitative variable that is not shown? *Shooting percentage*

Class Activity 1.2: Sample Surveys

Let us say that I am interested in current junior school students at MBA and would like more information about them. I could ask every junior school student a series of questions or I could select a group of junior school students who will then represent the entire junior school with their responses. In this study, all current junior school students make up my **population**, the entire group in which I have interest. If I tried to ask every single junior school student my questions that would be a **census**, an attempt to collect information from every member of a population. If I simply select a group of students then that group of students would be called my **sample**, the part of the population from which we collect information. Observational studies that gather data by sampling are called **sample surveys**.

As a class, let's think of an interesting yes/no question you could ask a junior school student.

Question: Do you use your locker on a daily basis?

Homework Assignment 1.1: Junior School Sample Survey

In the space below, record the names and responses from five MBA 7th graders and five MBA 8th graders to the class question above.

Name	Grade	Response
Jax Leasure	7	Yes
Jack Elrod	7	No
Collin Dunelli	8	No
Brittain Buzby	7	No
Sam	7	No
Asher	8	Yes
Gabe	8	Yes
Jack Elrod	8	No
Frank Johns	8	Yes
Ben Crawford	7	No

yes (bar)
7th   8th
 1   3

Yes total (pie) / No total
    4   6

Good Data is a product of intelligent human effort

Next, let's represent these data as a **pie chart** and a **bar graph** in the right column of the table below. A pie chart is a circular graphic that is divided into sections to illustrate parts of a whole. A bar graph compares different values by using rectangular bars with lengths proportional to the values they represent. For the pie chart, illustrate the percent of "Yes" and "No" responses you received in the junior school. For the bar graph, compare the percent of "Yes" responses you received from 7th graders to the percent of "Yes" responses you received from 8th graders.

Example: Is math your favorite class at MBA?	Do you use your locker everyday?
Pie chart: No 30%, Yes 70%	Pie chart (hand-drawn): NO 67% (6), YES 23% (4)
Bar graph: Percent who said "Yes" — 7th: 60, 8th: 80	Bar graph (hand-drawn): 7th: 20%, 8th: 60%

※ A pie chart lets you see parts of wholes
※ You're allowed to have 99.9% because of rounding off errors
※ A bar graph allows you to compare ~~quantitative scores/values~~ percents side by side

Class Activity 1.3: Comparing Survey Results

If you compare your data with your classmates' data, you are likely to see that results vary from survey to survey. It is not surprising that one sample of ten junior school students yielded different results from a sample of ten other students. Even with perfect sampling methods you will have **sampling variability**, variation between different samples from the same population. In addition, with no instruction, it is quite possible that the sampling method you used was **biased**. A biased sampling method is one that, in repeated trials, would consistently misrepresent the population in the same way. For example, if you asked 10 students who were walking out of Mathcounts practice, there is a good chance that a great proportion of that group would list math as their favorite class compared to the entire population of junior school students. If you walked into a meeting of MBA's Art Club, that group may consist of many students who would list Art as their favorite class and thus the number of students who say math is their favorite class would be lower than you would find in the entire population.

We will discuss bias and variability more in chapter 2, but typically a good sampling method has small bias and small variability.

To reduce bias, __use random sampling__.

To reduce the variability of a sample, __use a larger sample__.

Class Activity 1.4: Experimentation

Linus Pauling is the only person who has ever been awarded two unshared Nobel Prizes; he won the Nobel Prize in Chemistry in 1954 and the Nobel Peace Prize in 1962. In 1970, he published *Vitamin C and the Common Cold* in which he advocated for the use of high doses of Vitamin C to lower the frequency and duration of colds. An experiment to test this claim was performed in Toronto.

The *double-blind* experiment sought to enroll a *large number of subjects* (1000) in the study, though only 818 completed the entire study. Subjects were assigned *at random* to two groups. The experimental group received 1 gram of vitamin C a day and 4 grams at the first sign of a cold. The *control group* received a *placebo* that looked like the vitamin C capsules but had no active ingredient. The experiment wanted to see if vitamin C prevented colds from occurring and if it shortened the duration of colds if they did occur. The study found that there was a *statistically significant* difference between the experimental and control group both in terms of incidence (*P<.05*) and duration of colds (*P<.001*).[4]

The appeal of experimentation is clear from this example. Experiments can give evidence for cause and effect. In this case, evidence was provided that supported the hypothesis that high doses of vitamin C can help prevent and shorten colds.

Why was it important that the study used such a *large subject group*?

Larger subject groups reduce the possible variabilities

Why were subjects assigned to the control and experimental group *at random*?

In order to reduce bias

Why is it important to have a *control group* at all?

To have a reference

Why did the experimenters give the control group a *placebo*?

In order to produce the same psychological effects as those in the experimental group

What does it mean to be *double-blind* and why is it important?

Neither the participants nor the experimenters know who's actually getting treatment — it reduces bias

What does *statistically significant* mean? How is it related to *P<.05 and P<.001*?

There's a large difference between the statistics; P<.05 is much greater than P<.001

These questions help us to start thinking about experimental design and how best to find evidence for causal relationships. There is, however, much more to think about. How could people starting the study but not finishing it (dropouts) affect the reliability of results? When is ethical to perform experiments like this where half of the subjects involved are receiving a placebo with no active ingredient? Does it change if we are talking about a disease more severe than the common cold? Many scientists have argued against Pauling's claims. When scientists disagree, how do you decide which one to believe? We will discuss questions like these in chapter 3.

Homework Assignment 1.2: Chapter 1 Review

When sampling and experimenting, it is good practice to take a large sample and use a large group of subjects respectively. Why?

Large samples give a broader scope, and offer a more accurate representation of a given population as a result. It also reduces variability within the group.

What is the risk of not selecting a sample group at random? What about not assigning subjects to the control and experimental group at random?

By not choosing a group at random, one is innately biased. The same goes for when one actively handpicks who will be control/experimental.

Part A: Producing Data

Chapter 2. Sampling for Observational Studies

Class Activity 2.1: A Fireworks Display for MBA's Sesquicentennial?

MBA is excited about its sesquicentennial celebration in 2017 which marks 150 years since the school officially opened in 1867. MBA's administration wonders if a large fireworks display would be an appropriate way to mark the occasion. Unsure whether or not the community would be in favor of such a display, the administration asks a handful of statistics students to find out.

There is some proportion of the surrounding community that would support a fireworks display. That proportion is a **parameter**, a number that describes the population. The variable **p** is used to represent a population parameter. Unfortunately, it is impossible to know what the parameter value is unless you perform a census, and even then results are not always accurate. Since there are only a handful of students who are working on this project and they have a limited amount of time, it would be more realistic for them to take a sample of the population and then use the responses from that sample to determine a **statistic**, a number that describes a sample. The variable $\hat{p}$ ("p-hat") is used to represent a sample statistic. A well-drawn sample yields a statistic that can be useful in estimating the value of the population parameter.

The students decide to gather to set up a booth at the next home football game and poll people as they are heading to the stands. The students ask the first 40 people who arrive at the game if they would support a large fireworks display to celebrate MBA's sesquicentennial and are excited when 36 of the 40 people polled say that they would be in favor of the fireworks display.

Find $\hat{p}$ for this sample. _Yes - 90%_

$\hat{P}$ = sample statistic
P = Population perameter

Do you think this $\hat{p}$ is a good estimate of the population parameter p?

No, because it is only a sample taken from people attending an MBA event; it's biased and is a small sample size.

What the students did at the football game is called **convenience sampling**, selecting individuals for a sample that are the easiest to reach. While it certainly made polling painless for the students, convenience samples frequently result in bias, a concept introduced in chapter 1. A study is biased if it systematically favors certain results. In this example, it is quite likely that those people in attendance at an MBA football game are not representative of the community living around MBA. Specifically, it is a section of the population that is more likely to support MBA and MBA activities and thus a $\hat{p}$ generated using this method will likely be greater than the population parameter p.

The students scratch the results from the football game poll, and ask The Tennessean if they will put the following in its Lifestyle section:

> Montgomery Bell Academy is considering a fireworks display to celebrate its 150th anniversary. Please call (615) 555-3141 to share whether or not you would support such a display.

After a week, the students received 50 phone calls. 10 were in support of the fireworks display and 40 were opposed to the display.

Find $\hat{p}$ for this sample. *Yes = 25%*

Do you think this $\hat{p}$ is a good estimate of the population parameter p?

No; the Tennessean allows responses from any TN resident, and only old people read it.

This second attempt by the students was what is called a **voluntary response sample** (or self-selecting sample). It also frequently results in bias because those who choose to respond tend to be those who care enough about the issue to make the effort to do so. People who would support a fireworks display are less likely to call in to voice their support than those who oppose a fireworks display. As a result, this $\hat{p}$ is likely smaller than the population parameter p.

To avoid the bias that can be found in both convenience samples and voluntary response samples, we need a method that selects a sample using impersonal chance. The easiest way to do this is with a **simple random sample (SRS)**. A simple random sample selects n individuals in such a way that every set of n individuals that exists in the population has an equal chance of being selected. For example, if you wrote the name of every individual in a population on a piece of paper, put those pieces in a hat, shook it up, and drew n pieces out, that would be an SRS. If your population is a class of 15 students and you are trying to take an SRS of size 3, the hat will work just fine. If your population is thousands of Nashville residents, it might be more practical to use a **random number generator** or **table of random digits**.

Let's see how to select an SRS of 3 from our class using a random number generator and a table of random digits:

Using a Table of Random Digits

Step 1: Assign a number to every person in the class. If there are between 1 and 10 students in the class, use single digits (e.g. Tom = 0, Dick = 1, Harry = 2…). If there are between 11 and 100 students in the class, use two digits (e.g. Tom = 00, Dick = 01, Harry = 02…).

Name	Numerical Label

Step 2: A table of random digits is a long string of digits 0 through 9 that is generated in such a way that all 10 digits have an equal chance of being selected for each entry. Tables of random digits are often labeled by row. All rows are generated using the same method so any one can be used to select an SRS. Let's use line 132 (and 133 if necessary) taken from a table of random digits.

Line
...
132   66810 31421 47836 02603 15377 22725 14592 15412
133   41267 37741 04900 11776 47483 12513 95034 12630

If you labeled members of the class with single digits, read groups of single digits, i.e. 6, 6, 8, 1, 0, 3, 1, 4, 2, 1... If you labeled members of the class with two digits, read groups of two digits, i.e. 66, 81, 03, 14, 21, 47... If you read one of our students' labels, he is part of the SRS. If you read a number that has already been read or is not assigned to one of our students' labels, skip it.

First three unique numerical labels that corresponded to students:

_____ , _____ and _____ .

The three students in our SRS:
_____ , _____ and _____ .

Using Random Number Generator

Step 1: Assign a number to every person in the class. Since we have already done this when we selected our SRS with a Table of Random Digits, let's use those values (if you used two digits previously, it is okay now to simply think of your first student at 0, second student as 1 etc. instead of as 00 and 01).

Step 2: If you have a TI-83 or TI-84, press MATH, select PRB, and then 5: randInt(. Then complete the command as randInt(0,x) where x is the numerical label of the last student in the class. For example, if there are 14 students in the class labeled 0 through 13, your home screen should look like the following:

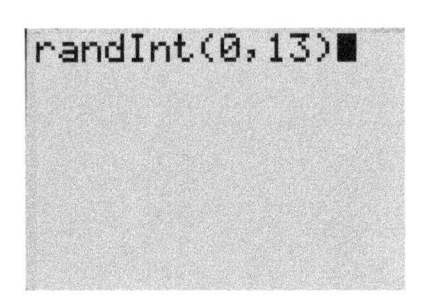

```
randInt(0,13)
```

If you press ENTER, the calculator will select a number between 0 and x. To produce more digits, continue pressing ENTER. Using this method, you do not have to worry about skipping numbers that are not assigned to an individual in your population like you did when using the Table of Random Digits. The randInt command may, however, give you the same value more than once. If a value repeats, skip it just like we did when selecting an SRS with a Table of Random Digits.

First three unique numerical labels:

_____ , _____ and _____ .

The three students in our SRS:

_____ , _____ and _____ .

Homework Assignment 2.1: Bias and Simple Random Samples

In the mid-1970s, Ann Landers responded to a young married couple who was debating whether or not to have children by asking her audience to write in with a response to the question: "If you had it to do over again, would you have children?" [5]

What kind of sampling method was Ann Landers using when she asked this question?
Convenience sampling

Do you think the sample statistic $\hat{p}$ of people who would do it all over again that Ann Landers received was higher or lower than the true population parameter p?
higher

If you were selecting an SRS of 50 students with all current MBA students as your population, how would you label the first student: 0, 00, 000, or 0000? *000*

To label the current MBA students for the SRS, a student uses an MBA phone directory. To save time, he simply selects a single random number and then uses that number and the 49 numbers that follow for his random sample (if he goes past the largest assigned number, he goes back to the first student and continues). The student justifies this process by saying that every student still has the same chance of being selected as any other student. This is true but this sample is not an SRS. Why is this sample not an SRS and how could this sampling method result in bias?
NOT EVERYBODY'S IN THE PHONEBOOK. The selection of the following 49 numbers is not because they were actually selected; it's a result of consequence by the first selected number. Not a true SRS. It's not random for the following 49, making it innately biased.

Even though voluntary response samples can result in bias, they are still very commonly used. Can you give an example of a website that uses voluntary response sampling?
Yelp! Radioshows

Using your own TI-83 or TI-84 calculator, or borrowing one from a friend, type in 2016, then press STO→, MATH, choose PRB and select 1:rand. Press ENTER to complete the command. Your screen should look like the following:

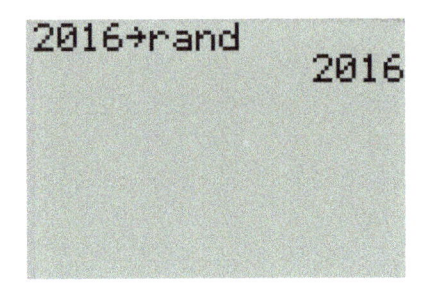

```
2016→rand
              2016
```

Now do the command randInt(1,100) and write down your result: _____. In our next class, we will discuss this result and what it tells us about the nature of randomness.

Class Activity 2.2: Random Numbers and Pseudo-Random Numbers

Compare the result you received on the last 2.1 homework problem to your peers' results. If you were using the same make of calculator, you should have received the same answer. If you were using a TI-83 or TI-84, you should have received a 6. If you continued pressing ENTER, you should have received 50 next and then 33. Computer programs and calculators are unable to produce truly random numbers. The best they can do is produce numbers that appear random, called **pseudo-random numbers**. When you inputted 2016→rand on your calculator, you were selecting a seed, or starting value, for your random digit generation. Since all TI-83 and TI-84 calculators have the same random integer generator program, if you set two calculators to the same seed value, they will output the same "random" results. Since these numbers are not truly random, there can be bad consequences when using them for more advanced statistical analysis, but for our purposes they are still effective at statistical randomizing and can help reduce the bias that can be present with alternative sampling methods.

Class Activity 2.3: Sample Size and Variability

The students researching fireworks have seen that using an SRS can help reduce the bias in their results. Now they need to define their population. The group decides that the population whose fireworks opinion they need to worry about is those who live within the 37205 zip code. This makes for a population of about 25,000 people. The group selects an SRS of 16 people and asks those 16 people about the fireworks display. 9 of the 16 say that they would be in favor of the fireworks display.

Find $\hat{p}$ for this sample. _____9/16_____

By using an SRS, the group has made progress in terms of minimizing the bias in their sampling method, but there is another potential issue, sampling variability. Sampling variability was mentioned in chapter 1 and describes how spread out sampling results (the $\hat{p}$ values) would be in repeated samples and thus how confident we can be that a $\hat{p}$ is a good estimate of the population parameter p.

A sampling method should try to have small bias and small variability. We attempted to have small bias by using random sampling. To minimize variability, we increase our sample size. Let's explore the relationship between variability and sample size.

To see how p values and $\hat{p}$ values are related for different sample sizes, let's assume that we know the parameter value p. For this example, let's say that p = .6 (meaning that 60% of those who live in the 37205 zip code would support the fireworks display). To make the math easy, let's say that we labeled those who supported the fireworks display with numbers 1 through 15,000 and those who oppose the fireworks display as numbers 15,001 through 25,000. Now let's simulate selecting a sample of 16 individuals from this population.

Selecting a sample of 16 individuals from a population of 25,000

Step 1: To make sure that all the students in the class aren't set to the same seed value on their random integer generators, type the last four digits of your phone number into your calculator, then press STO→, MATH, and then select PRB and 1:rand and then hit ENTER. For example, if the last four digits of your home phone number were 2718 your screen should look like the following:

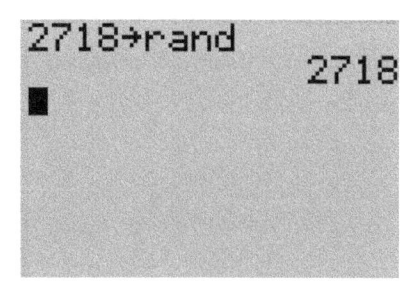

```
2718→rand
            2718
■
```

Step 2: To simulate the selection of 16 individuals from a population of 25,000, type in randInt(1,25000,16). Typing ",16" at the end of the command tells the calculator to select 16 numbers between 1 and 25,000. It is unlikely but possible that there will be a repeat in the 16 numbers. If you are worried about a repeat result, you can type randInt(1,25000,17) so you will still have 16 unique numbers even if there is a repeated number. After performing this command, record your 16 values in the middle column in the table below. In the right column, indicate whether the person would say he/she is in favor of the fireworks (if the number is between 1 and 15,000 inclusive) or opposed to the fireworks (if the number is between 15,001 and 25,000 inclusive)

Individual	Number	In favor of or opposed to fireworks
1		
2		
3		
4		
5		
6		
7		
8		
9		
10		
11		
12		
13		
14		
15		
16		

Find $\hat{p}$ for this sample. _____

Record the $\hat{p}$ of your classmates below to see how effective a sample size of 16 was at estimating the population parameter of .6:

Student #1	$\hat{p}=$	Student #9	$\hat{p}=$
Student #2	$\hat{p}=$	Student #10	$\hat{p}=$
Student #3	$\hat{p}=$	Student #11	$\hat{p}=$
Student #4	$\hat{p}=$	Student #12	$\hat{p}=$
Student #5	$\hat{p}=$	Student #13	$\hat{p}=$
Student #6	$\hat{p}=$	Student #14	$\hat{p}=$
Student #7	$\hat{p}=$	Student #15	$\hat{p}=$
Student #8	$\hat{p}=$	Student #16	$\hat{p}=$

Homework Assignment 2.2: Variability with a Larger Sample Size

For homework, let's repeat the exercise we did in class, but this time, let's use a sample size of 100. Use randInt(1,25000,103) to be confident you will have 100 unique results. Record your results in the table below and then find $\hat{p}$ for your sample:

Individual	Number	In favor of or opposed to fireworks
1		
2		
3		
4		
5		
6		
7		
8		
9		
10		
11		
12		
13		
14		
15		
16		
17		
18		
19		
20		
21		
22		
23		
24		
25		
26		
27		
28		
29		
30		
31		
32		
33		
34		
35		

36		
37		
38		
39		
40		
41		
42		
43		
44		
45		
46		
47		
48		
49		
50		
51		
52		
53		
54		
55		
56		
57		
58		
59		
60		
61		
62		
63		
64		
65		
66		
67		
68		
69		
70		
71		
72		
73		
74		
75		
76		

77		
78		
79		
80		
81		
82		
83		
84		
85		
86		
87		
88		
89		
90		
91		
92		
93		
94		
95		
96		
97		
98		
99		
100		

Find $\hat{p}$ for this sample. _____

Class Activity 2.4: A Quick Method for Margin of Error

Record the $\hat{p}$ of your classmates below to see how effective a sample size of 100 was at estimating the population parameter of .6:

Student #1	$\hat{p}=$	Student #9	$\hat{p}=$
Student #2	$\hat{p}=$	Student #10	$\hat{p}=$
Student #3	$\hat{p}=$	Student #11	$\hat{p}=$
Student #4	$\hat{p}=$	Student #12	$\hat{p}=$
Student #5	$\hat{p}=$	Student #13	$\hat{p}=$
Student #6	$\hat{p}=$	Student #14	$\hat{p}=$
Student #7	$\hat{p}=$	Student #15	$\hat{p}=$
Student #8	$\hat{p}=$	Student #16	$\hat{p}=$

When trying to express how confident we are that $\hat{p}$ is close to our p value, we use a **confidence statement**. A confidence statement includes both a **margin of error** and a **level of confidence**. For example, a confidence statement might say "we can say with 90% confidence that the p is equal to $\hat{p} \pm 5\%$." In that example, the level of confidence is 90% and the margin of error is 5%. Let's say that our $\hat{p}$ was .34. That confidence statement is saying that 90% of the time, the population parameter p is between .34± 5%. That is, we can be 90% confident that p is between .29 and .39.

The most common level of confidence used is 95%. There is a quick formula for estimating margin of error for 95% confidence. Our **quick method for margin of error** tells us that when we draw a simple random sample of size n, the margin of error for 95% confidence is approximately equal to $\frac{1}{\sqrt{n}}$.

That means that when our sample size n was 16, our margin of error for 95% confidence was approximately equal to:

$$\frac{1}{\sqrt{n}} = \frac{1}{\sqrt{16}} = \frac{1}{4} = .25$$

That also means that when our sample size n was 100, our margin of error for 95% confidence was approximately equal to:

$$\frac{1}{\sqrt{n}} = \frac{1}{\sqrt{100}} = \frac{1}{10} = .1$$

That means that 95% of the time we expect the population parameter p to be within .25 of our $\hat{p}$ when our sample size n is 16 and 95% of the time we expect our population parameter p to be within .1 of our $\hat{p}$ when our sample size n is 100. Let's see how often p was within the margin of error in our class when our sample size n was 16. Go back to Class Activity 2.3 and copy down the $\hat{p}$ values from the class. Then create a range between which we are 95% confident we will capture the p value by adding and subtracting .25 from $\hat{p}$. Finally, indicate whether or not p (.6) was captured within the margin of error.

Conclusion always applies to population, not the sample
Conclusion is never completely certain
A sample survey can use confidence level of other than standard 95%
Don't usually report margin of error unless you use 95% confidence
Smaller margin of error with consistent confidence requires a larger sample

Student	$\hat{p}$	$\hat{p} \pm .25$	Was p (.6) within margin of error?
1			
2			
3			
4			
5			
6			
7			
8			
9			
10			
11			
12			
13			
14			
15			
16			

In what percent of our samples did we capture the truth about the population (p) within our margin of error? _____

Is that result surprising? _____

_____

_____

Now let's see how often p was within the margin of error in our class when our sample size n was 100. Go back to Class Activity 2.4 and copy down the $\hat{p}$ values from the class. Then create a range between which we are 95% confident we will capture the p value by adding and subtracting .1 from $\hat{p}$. Finally, indicate whether or not p (.6) was captured within the margin of error.

Student	$\hat{p}$	$\hat{p} \pm .1$	Was p within margin of error?
1			
2			
3			
4			
5			
6			
7			
8			
9			
10			
11			
12			
13			
14			
15			
16			

In what percent of our samples did we capture the truth about the population (p) within our margin of error? _____

Is that result surprising? _____

_____

_____

Homework Assignment 2.3: Working With Margin of Error

We found using the quick method for margin of error that it takes a sample size of 100 to have a margin of error of 10% with a confidence level of 95%. How large would your sample size have to be in order to have a margin of error of 5%? n = __400__

In order to halve your margin of error, by how much did you need to multiply your sample size? __4__

Use the quick method for margin of error to find how large a sample would have to be in order to have 95% confidence that $\hat{p} \pm 2\%$ would capture the population parameter p. n = __2500__

Class Activity 2.5: Population Size vs. Sample Size

Let's say the school wants a large enough sample size in order to have 95% confidence with a margin of error of 3%. Our quick method for margin of error shows us that n should equal 1,112 (rounding up from $1,111.\overline{1}$). The students set out to do another SRS of those people who live in the 37205 zip code, but this time with this larger sample size. Right before we start selecting our sample, however, the administration sees that we are only using those who live in the 37205 zip code in our **sampling frame**, the list from which we draw our sample. The administration says that this will be a huge fireworks display that will be heard and seen by people all throughout Nashville. That means that our population size, denoted by a capital N, just went from about 25,000 to over 600,000.

The students are nervous that this means they will need a larger sample, but the good news is they won't. As long as the population is much larger than the sample size, the variability of the statistic does not depend on the size of the population. As a rule of thumb, the population size does not have to be considered in a margin of error calculation as long as the population N is at least 10 times as large as the sample size n.

Finally, the students are able to take their SRS of size 1,112 in Nashville and 589 of the 1,112 say that they are in favor of a fireworks display.

Find $\hat{p}$ for this sample (round to the nearest hundredths). __0.53__

Make a confidence statement about Nashville and its residents' feelings about a potential fireworks display. __We are 95% that the true proportion, P, of people who want the fireworks display lie in the interval of 0.53 to 0.56__

Is it possible that the majority of Nashville residents oppose the fireworks display despite the students' sample statistic? _____

Is it probable? _____

So far we have only been able to make confidence statements with a confidence level of 95%. In chapter 9, you will learn how to make a confidence statement with any confidence level you want.

Class Activity 2.6: Alternatives to Simple Random Samples

We might assume that different groups within the Nashville population would be more or less likely to support a fireworks display. For example, we might think that younger members of the community might be more likely to support a fireworks display and the elderly might be more likely to go to bed early and get frustrated with the noise that comes with a fireworks display.

If there are groups, or strata, within a population that are distinct in some way that is significant for the question at hand, we can use a method of sampling known as **stratified random sampling**. In stratified random sampling the population is broken up into groups, or strata, and then an SRS is taken from each stratum. A stratified random sample can help us ensure that every age group is represented proportionally in our final results. For example, in Nashville, approximately 22% of the population is under 18 years old, 68% is between 18 and 64, and approximately 10% is 65 and older.[6] If we use an SRS it is possible that 23% of our sample will be under 18 years old, 70% will be between 18 and 64 and 7% will be 65 and older. Or, we can study the strata separately or intentionally represent one or more strata disproportionately. For example, MBA might care more about the opinion of those under 18 because it wants the youth of Nashville to be excited about the display to increase applications and thus select a higher proportion of its total sample from the under 18 group than one would find on average in an SRS.

The last sampling method we will discuss is **cluster sampling**. In a cluster sample, the population is first broken up into groups, or clusters, and then a simple random sample is taken to select some of these clusters. Then a simple random sample is performed within the selected clusters to form our sample. A cluster sample varies from a stratified random sample in a couple ways. First, unlike a particular stratum in stratified random samples that is supposed to contain similar individuals, clusters are selected in the hopes that they will represent the diversity found in the entire population. Second, in cluster sampling you do not need to sample from every cluster. The hope is that taking a random sample from the randomly selected clusters will provide a sample representative of the population. One could envision doing cluster sampling with our fireworks example by breaking Nashville up into neighborhoods, randomly selecting a few of those neighborhoods, and then randomly selecting some members of those neighborhoods to be a part of the sample. The major benefit of cluster sampling is that it is economical. Imagine trying to apply a label to every member of the population of Nashville (over 600,000 individuals) and then try to reach the over 1,000 people randomly selected for our sample. It would be much easier to break Nashville into neighborhoods, randomly select some, and then draw samples from those. The major downside is that the clusters may not be representative of the population as a whole and by selecting non-representative clusters, we may introduce bias into our sample.

All good samples are **probability samples**, samples that involve chance selection in at least one step. Simple random samples, stratified random samples and cluster samples are probability samples but convenience samples and voluntary response samples are not.

Homework Assignment 2.4: Stratified Random Samples and Cluster Samples

Let's say I wanted to figure out how many MBA students used the indoor rowing machines (ergs). I could do a simple random sample that selected 50 students from the school and asked them the question "Have you ever used MBA's indoor rowing machine?" It might make more sense to consider two strata, members of MBA's crew team and non-members. Give an example of other strata into which I could divide the MBA population before taking an SRS from each stratum.

Highschoolers vs. Junior Schoolers

I want to find out whether students would be in favor of or opposed to changing MBA to a block schedule. I decide the easiest way to get results is by doing a cluster sample with advisories as my clusters. I use an SRS to select 10 advisories and then select an SRS of 5 students within each advisory to get my total sample of 50 students. If I do not believe that this method will result in more bias than an SRS of the entire MBA population, what assumption about the advisories at MBA has to be true?

The advisories contain identical amounts of people within each grade in relation to the others. There's at least 5 students in each advisory

"resistant"

Mean is not resistant

A _____ can drastically

Class Activity 2.7: Sampling Errors

There are two types of errors that are present in sampling, **sampling errors** and **nonsampling errors**. Sampling errors result from the act of taking a sample. Nonsampling errors are those errors that do not result from the act of taking a sample and can be present even in a census.

The first type of sampling error, **random sampling error**, is the result of sampling variability or the fact that $\hat{p}$ values are imperfect in trying to estimate the population parameter value p. The only error accounted for in a margin of error is random sampling error. You can see that this "error" is not an error in the sense that you normally use the word. Even a perfectly conducted sample survey would have random sampling error. It does not mean that a mistake was made. It simply means that sample statistic values deviate from the population parameter.

The second type of sampling error, **undercoverage**, occurs when the population is not equal to the sampling frame. For example, if the students polled people about fireworks over the phone, there is no way a person without a phone could be included in the sample. This is especially significant when those left out of the sample would respond differently than those who are included in the sample.

As stated before, nonsampling errors can occur even when a census is performed. The three major types of nonsampling errors are **processing errors**, **response errors** and **nonresponse**. Processing errors are mistakes performed while doing mechanical tasks like data entry on a computer or arithmetic. While processing errors are less common now that more and more calculation is done by computers, they will exist as long as a human element is present in the sampling process. Response errors are simply when an incorrect response is given. This can be unintentional, for example if someone forgets how many times he has been to the grocery store in the past month or intentional, if someone says that he weighs 20 pounds less than he actually does because he does not want to share his true weight. While not technically an error, people may also give different answers to the same question depending on how it is worded. When people are asked if they believe in the government providing "assistance to the poor" they tend to be much more supportive than they are of "welfare" programs, even though they are the same. A higher proportion of people might say they are in favor of a "fireworks display free to the public that celebrates secondary education and Nashville's history" than a "massive late-night fireworks display to celebrate a century and a half of excellence at Montgomery Bell Academy." Lastly, nonresponse is the failure to obtain data from an individual selected to be part of the sample. Much like undercoverage, this can create bias if those individuals from whom we do not obtain data are different in some way relating to the survey question. If people hang up the phone as soon as they hear the words MBA, it is possible that a smaller proportion of those who do not respond would be in favor of a fireworks display than those who do.

While much of this chapter has been spent focusing on the difficulties that statisticians face when using sample surveys, it is important to note that sampling, when done carefully, has proven to be a very effective way to gauge the feelings of a population without having to poll every member of the population. By making sure to avoid bias, using large sample sizes and reporting margin of error and confidence level, polling agencies are able to give us great insights into large populations.

Homework Assignment 2.5: Sampling Errors in Review

One of the most famous blunders in sampling was the presidential poll done in 1936 by *The Literary Digest*. Republican Alf Landon was running against Democrat Franklin Roosevelt and *The Literary Digest*, confident after correctly predicting the previous five presidential elections, sent out a massive 10 million ballots to its subscribers and to automobile and telephone owners whose information could be found in public records. *The Literary Digest's* prediction: Alf Landon will win 57% of the popular vote. What actually happened? Franklin Delano Roosevelt took over 60% of the popular vote and the electoral votes of an incredible 46 out of 48 states (Hawaii and Alaska were not yet part of the Union).[7] *The Literary Digest* certainly had a large enough sample size so the issue was not variability. Since only those who subscribed to *The Literary Digest* and those who owned a phone or automobile were included in the sampling frame and *The Literary Digest* was trying to predict the popular vote of all eligible voters in the United States, what type of sampling error was present in this poll? _____

The Literary Digest sent out 10 million ballots but received only about 2.4 million responses.[8] This is an example of what type of nonsampling error?

_____

There remains some disagreement about what was responsible for the great difference between *The Literary Digest*'s prediction and what happened on election day. How do you think the two errors you just mentioned could have added bias to the prediction? Keep in mind what was happening in the US in 1936.

_____

_____

_____

In 2014, the Tennessee gubernatorial election saw incumbent Republican Bill Haslam face off against Charles "Charlie" Brown. On August 11-12, 2014 Rasmussen Reports polled 750 likely voters and 55% said they were going to vote for Bill Haslam.[9]

The stated margin of error for the poll was 3% with 95% level of confidence.[10] What margin of error do you get when you use the quick method for margin of error? _____

On November 4, 2014, Bill Haslam won the election with 70.3% of the vote. This is well outside the margin of error on Rasmussen's poll. The margin of error only accounts for random sampling error. It is unlikely that random sampling error is completely responsible for the difference between Rasmussen's 55% and the 70.3% that occurred on Election Day. What else do you think could be responsible for this change? Keep in mind that this poll was taken almost three months before the election.

_____

_____

_____

_____

Chapter 3. Experimenting

Class Activity 3.1: A Headache Outbreak at MBA

There is an epidemic of headaches at MBA! Students are in pain, staying home from school, and when they are in school they are having trouble focusing on class material without aggravating their condition. Mrs. Power tells Mr. Gioia that she has the cure, an herbal tea that her family has used for generations to help with headaches.

There are two types of data collection, passive and active. Observational studies use passive data collection. This means that the subjects being studied are not intentionally influenced. **Experiments** involve active data production where a **treatment**, or condition, is imposed on a group of **subjects**, the individuals being studied. In order to see if Mrs. Power's herbal tea can help with headaches, we will need to perform an experiment. Our subjects will be MBA students and the treatment applied will be having the students drink the herbal tea.

When we look at the results of the herbal tea study, the **explanatory variable**, also referred to as the independent variable, will be whether the student drank the tea and if we get more advanced, how much and how often the tea was consumed. The **response variable**, also referred to as the dependent variable, will be the change in the health of the student, i.e. whether the headaches went away and if we get more advanced, the change in the severity of the headaches.

To test the herbal tea, the administration sends out an email to all six classes that says the following, "If you would like to be part of a study on headache reduction, please stop by Ball this afternoon immediately after 8th period." The administration is happy that 5 students show up to Ball immediately after school. The administrators have the 5 students drink the Power Family herbal tea, take the afternoon off from sports to rest and tell the students to come back the next day to report how they are feeling. The next day the 5 students come back, three of the five say they still have a headache and two of the five say they do not. A clever Statistics student is walking through the Ball building and overhears the details of the study. He is concerned about the study's design.

The first concern the student raises is that the results of the study may be **confounded** by a **lurking variable**. Two variables are confounded when their effects on a response variable cannot be distinguished. A lurking variable is a variable that has a significant effect on the relationship between variables in a study but is not included as one of the explanatory variables. In this study, two variables were intentionally changed. The students drank the herbal tea and they took the afternoon off to rest. If the students show improvement, we will not be able to decide whether it was the tea or the additional rest that was responsible. Figure 3.1 illustrates this confounding. The arrows represent *causation*. The question mark is there to show that we cannot determine whether one or both of the variables are responsible if there is a reduction in headaches because they occur together in this experiment. In order to determine whether the tea is effective, the experiment should do its best to change only a single variable. In this case, the one variable should be the consumption of herbal tea.

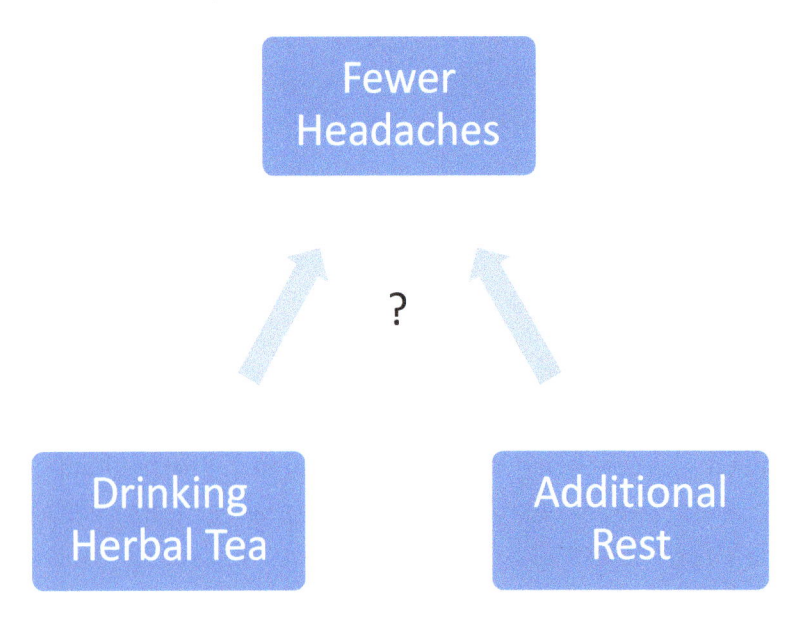

Figure 3.1

The student points out a second problem to the administration, the **placebo effect**. A placebo is a dummy treatment with no active ingredient. The placebo effect is the positive response that patients show to any treatment, even if the treatment is a placebo. It is a psychological effect combined with the fact that many patients will also improve for reasons other than the treatment. If students' headaches improve after drinking the tea, we need to find a way to determine whether the tea was responsible or whether this was just another example of the placebo effect. To do this, we add another group to our study, a group that receives a placebo treatment. In this situation the placebo would be something harmless that tasted like the herbal tea, such as flavored water. The assignment of students to the group that receives the Power Family herbal tea and the group that receives the placebo treatment would again be random to avoid any sort of bias in selection, e.g. if students were suffering more from headaches it would be tempting to keep them out of the placebo group, but that would compromise the data collected in the study. Since the group receiving the placebo allows us to control for lurking variables like the placebo effect in the study, it is referred to as the **control group**. If the assignment of subjects to the control group and to the group or groups receiving the treatment being tested is done at random the study is called a **randomized comparative experiment**.

The final problem that the student identifies is that there are not nearly enough subjects in the study. Just like increasing our sample size in chapter 2 reduced the risk of chance variation, increasing the number of subjects in each treatment group reduces the risk of chance variation. To understand this more clearly, let's suppose that 80% of those students who drink the tea see headaches disappear completely within a week and only 40% of those who receive the placebo see headaches disappear completely within a week. If there are 5 students in each group, this result could simply be the result of chance much like it would be very possible 4 out of the 5 students in the tea group to flip heads on a fair coin and only 2 out of the 5 students in the control group flip heads on a fair coin. If there are 500 students, it is incredibly unlikely that the 80% to 40% difference between the two groups would be the result of mere chance, much like it would be shocking for 400 of the 500 students in the tea group to flip heads on a fair coin while 200 of the 500 students in the control group to flip heads on a fair coin.

Homework Assignment 3.1: Redesigning the Herbal Tea Experiment

There were three major problems that the student identified in the administration's initial tea experiment. 1, the study did not control the effects of lurking variables like additional rest and the placebo effect. 2, the study did not use random selection in choosing treatment group(s). 3, the study only used five students. Explain the risk of all three of these design flaws and explain how you would redesign the herbal tea experiment to address these three issues.

1. Allows skewed/inaccurate info; Add a control group

2. Generates bias; Choose groups off of an SRS

3. Too small of a sample size; Make it 30 people

THE KEY TO ANY EXPERIMENT IS RANDOM SAMPLING

Clinical trials are experiments like this one that test the effectiveness of different medical treatments on humans. In addition to being randomized comparative experiments, whenever it is possible the experiments are also **double-blind**. In a double-blind study neither the subjects nor the people working with the subjects know who is receiving which treatment. What would be the risk of having an experiment, like this herbal tea experiment, that is not double-blind? Describe how you would make this herbal tea experiment double-blind in practice.

_____
_____
_____
_____
_____

Class Activity 3.2: Other Issues with Clinical Trials

In chapter 2, we discussed a number of sampling and nonsampling errors that can be responsible for our sample statistic deviating from the population parameter. Even the best designed experiments can encounter similar issues. Much like surveys suffer from nonresponse, experiments have to deal with the presence of **refusals**, people who are unwilling to participate in the study. Even if people agree to be subjects in an experiment, the researchers still need to worry about **nonadherers**, subjects who do not follow experimental treatment instructions and **dropouts**, subjects who begin but do not finish the experiment.

If refusals occurred randomly when subjects were being selected for the experiment, the researchers would simply need to ensure that enough people were asked to participate so that the final treatment group sizes were large enough to minimize chance variation. There is, however, a more serious risk that refusals pose.

When would the presence of refusals potentially bias the experimental results?
_____

What is an example of why a nonadherer would voluntarily participate in a clinical trial and then intentionally going against the experimental treatment and as a result potentially bias the results? _In order to benefit from the pay easily_

If people drop out of the study for reasons unrelated to the clinical trial, it is unlikely to introduce bias to the experimental data. Give an example of why subjects might drop out of an experiment that could bias the results. _Maybe a subject got a job in another city & can't participate any longer_

Class Activity 3.3: Statistical vs. Practical Significance

MBA performs a double-blind randomized comparative study to see if the Power family herbal tea is effective at treating headaches. The experiment finds that headaches went away faster in the tea treatment group than they did in the control group. Does this mean we should conclude that the tea is an effective tool for stopping headaches and that we should institute a mandatory tea time during break periods to curb headaches at MBA? It depends. The first question we have to ask ourselves is if the result was **statistically significant**. If an observed difference is so great that it would rarely occur by chance alone, the difference is said to be statistically significant.

Statistical significance is what statisticians are tasked with determining when performing significance tests (the subject of chapter 10). The general population, however, does not just care about statistical significance. The general population also cares about **practical significance**. A result is practically significant if the difference is statistically significant *and* is useful in an applied context. For example, we might find that students who drank the herbal tea on average took 6 days to recover from their headaches and students in the control group took 7 days to recover. It is possible that this is a statistically significant result which means that we can be confident that the herbal tea really does reduce the time necessary to recover from a headache. It is also possible that, despite being a statistically significant result, that we still lack practical significance. If students have to drink tea daily in order to only lower their recovery time on average by one day, students who do not enjoy tea may opt to just endure the headaches for a slightly longer time period. Practical significance is subjective and at times statisticians will stop after determining whether or not a result is statistically significant, but it is important to note that the two are not equivalent.

Homework Assignment 3.2: Types of Significance

Let's say that you have invented some new product that is supposed to improve upon a product already on the market. For example, you have invented a golf club that will help you hit farther or you have invented a new type of tennis string that lasts longer before breaking. Choose your new product and identify what makes it superior to previous products of the same type.

New Product: _Super shoes_

Special Quality: _Propels people on air_

Of course we cannot just take your word for it. Your new product will be tested against old products to see if it really is an improvement. Describe the data (number of products tested, results of test, etc.) you could produce in your experiment that would lead you to conclude the following.

Results are not statistically significant:
It looks like a regular docsider shoe

Results are statistically significant but not practically significant:
It works as a jet boosting shoe but is against the MBA dress code

Results are both statistically significant and practically significant:
It fits comfortably and secures the foot in regulation leather, while still being able to zoom across the quad

Class Activity 3.4: Other Experimental Designs

Thus far we have only discussed experiments with two treatment groups: the control group that receives a placebo and the experimental group that receives the new treatment being tested. Often there are more than two treatment groups when different dosages are tested. For example, in our herbal tea study there might be a placebo group, a group that drinks one cup of tea a day and a third group that drinks two cups of tea a day. We also have a larger number of treatment groups when we add additional explanatory variables because of interaction between the variables. For example, let's say that MBA thinks that herbal tea might help with headaches but that students also need to be getting the right number of hours of sleep. MBA could adjust the study to look at both students drinking 0, 1 or 2 cups of herbal tea a day and with students sleeping 6, 8, or 10 hours a night. For this study, subjects would be randomly assigned to one of 9 treatment groups. See Figure 3.2.

		Variable B Hours of Sleep		
		6 hours of sleep	8 hours of sleep	10 hours of sleep
Variable A	0 Cups of Tea	*Treatment 1*	*Treatment 2*	*Treatment 3*
Cups of	1 Cup of Tea	*Treatment 4*	*Treatment 5*	*Treatment 6*
Tea/Day	2 Cups of Tea	*Treatment 7*	*Treatment 8*	*Treatment 9*

Figure 3.2

Completely randomized designs are simple and with large subject groups chosen at random, researchers can be reasonably confident that the treatment groups will be similar. If we have more information about our subjects, however, there is other experimental design that can yield more precise results. One simple example is a **matched pairs design**. In a matched pairs design, pairs of subjects that are as similar as possible are selected and then they are assigned to the two treatment groups at random.

For example, let's say that we are testing a new running shoe and want to see how it affects mile times. I could do a randomized comparative experiment with the track team, but by random chance I may end up with a stronger group of runners in one of my treatment groups. If I know who the fastest runners are on the team, I can pair up the 1st and 2nd fastest runners, select one for the control group (some shoes already on the market) and one to try out the new shoes with a coin toss, random number generator, odds or evens in a table of random digits etc. Then I could do the same thing for the 3rd and 4th fastest runners etc. In some situations, you can even use the same person for both treatment groups. For example, I could have the fastest runner run a mile with the new shoes and then with an older model. I have to be careful that there is time between the runs and that exhaustion or other factors do not influence performance, but if I do this there is no better matched pair than a person to him or herself. If the researchers do a good job identifying the important variables in a study like this, a matched pairs design can yield more helpful results than a completely randomized design. On the other hand, if researchers miss an important lurking variable and thus match pairs of subjects who are not that similar and are likely to have very different results in the experiment, results can be misleading and it would have been better to use the simpler completely randomized design.

Matched pairs designs are actually just one type of **block design**. A block design breaks experimental subjects into blocks, groups that are known to be similar in a way relevant to the results of the study, before random assignment to treatment is carried out. This should sound very similar to the stratified random sampling we discussed in chapter 2. Block design is to experiments as stratified random sampling is to sampling. The benefits are the same as well. We can draw separate conclusions about each block as we will be assured an appropriately large number of subjects within each block will be given each treatment.

Let's say that Belmont University is working on some new promotional videos to show at college fairs. It is reasonable to expect that boys and girls may react differently to videos and Belmont may find that some would be better to bring to college fairs at boys' schools like MBA and others might be more effective at girls' schools like Harpeth Hall. To see which videos are most effective within each group, a block design like the one outlined in Figure 3.3 could be used.

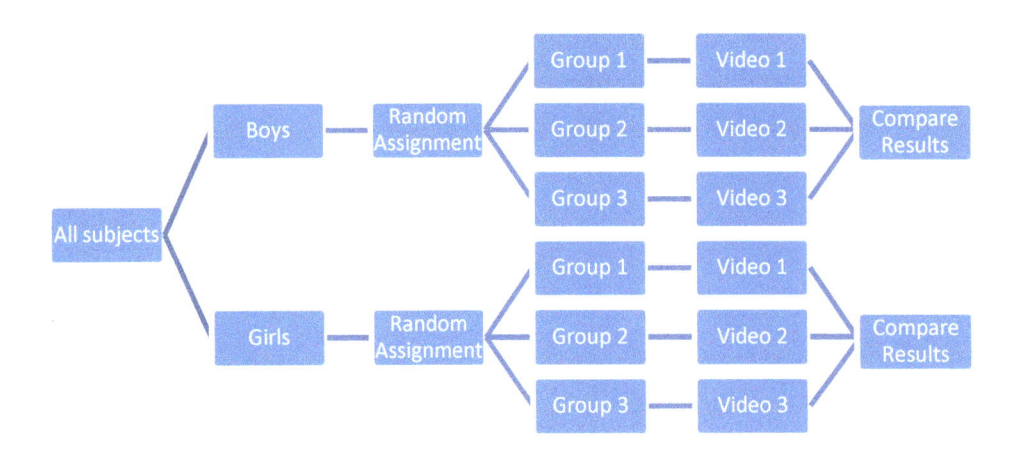

Figure 3.3

A key component in completely randomized designs and block designs is the attempt to control the effects of lurking variables by assigning subjects to treatment groups at random and trying to minimize the presence and/or effects of other variables. While this is great in theory, it can prove to be challenging and expensive in practice.

Consider Tennessee's Project STAR, which began in 1985 under former governor Lamar Alexander. STAR stands for Student/Teacher Achievement Ratio and attempted to find out if class size has an effect on student performance. An observational study would be unhelpful because schools that have smaller class sizes also tend to be different for other reasons. Schools with more teachers per student generally have more resources and provide more opportunities for students. As a result, those variables are confounded when trying to determine the effects of smaller class size. Project STAR randomly assigned students in 79 schools to one of three groups: a small class (13-17 students), a regular class (22-25 students) or a regular class with an aide in addition the regular teacher. Conventional wisdom says that smaller class size is better for students so, not surprisingly, some parents were able to successfully lobby to have their children moved into smaller classes and other students dropped out of the program. However, Project STAR did report results that were both statistically and practically significant. Students in small classes performed about .15 standard deviations better on standardized tests than those in regular sized classes. We will discuss what standard deviation means in the next chapter. California used Project STAR to partially justify its state-wide policy of class-size reduction and Lamar Alexander went on to be President George H.W. Bush's Secretary of Education. Tennessee's Project STAR also cost approximately $12 million. [11]

Even if you have the funds necessary to perform an experiment, it can be impossible to do experiments for ethical or practical reasons. In Charles Wheelan's "Naked Statistics" he titled the final chapter "Will going to Harvard change your life?"[12] Salary is only one way, and many would argue not a very good way, of evaluating success, but it is a popular metric. Wheelan noted that graduates of elite universities like Dartmouth, Princeton and Harvard have considerably higher median salaries 10 to 20 years after graduation than graduates of other colleges, but that information alone is not nearly enough to conclude that the experience at those colleges is more valuable. Students who attend those schools may simply be talented when they apply and that may be the reason they are accepted. It should be clear that it would not be possible to perform a randomized comparative experiment to test the value of an education at Harvard. Most students would not care to be placed at a particular college at random and colleges would also be opposed to their next incoming class being selected randomly.

Unable to perform an experiment, the best we can hope to do is to use the observational data that we have available. Economists Stacy Dale and Alan Krueger attempted to answer the question by addressing the most concerning lurking variables in this study: natural talent and work ethic. The same natural talent and work ethic that might be responsible for earning a student admission to an elite university may also be responsible for his or her success later in life. Dale and Krueger compared students who attended elite universities to those who did not attend elite universities *but were accepted*. This way we are comparing two groups of individuals all of whom were talented enough to earn admission to an elite university but only one group had the experience of attending an elite university. Dale and Krueger claimed that there was not a practically significant difference in earnings between those who attended elite universities and those were offered admission to elite universities and chose not to attend.[13]

Homework Assignment 3.3: Testing the Effect of an MBA Education

MBA graduates have a long history of success after graduation but a researcher could ask the same question of MBA that Wheelan, Dale and Krueger asked of Harvard. Is it the education provided at the school or simply the quality of the student accepted to the school that is responsible for students' success later in life? Like the Harvard study, it would not be practically possible to do a randomized comparative experiment to test the effect of an MBA education. Why is that?

Because your population is innately biased because they're from the school

If you used the same approach that Dale and Krueger used to test the effect of a Harvard education to test an MBA education, what two groups would you compare?

Harvard grads & non-Harvard grads but people who were accepted

The hope in comparing these two groups is that the talent and work ethic necessary to gain admission to MBA would be controlled because both groups were offered admission. The problem is that other lurking variables still may exist. List a few other variables that could influence both the likelihood that a student would choose to matriculate at MBA after being offered admission *and* could also have an effect on that student's ability to be successful in the future.

Family connections, financial stability, race, gender, proximity to home

To control the effects of these other variables, researchers will sometimes use the technique of **matching**, or attempting to find pairs of subjects who have similar observable characteristics except for the one being studied. It can be challenging to find enough similar subjects to draw any useful conclusions, which means that researchers are often forced to attempt to adjust for confounding variables. That is, if the effect of a confounding variable on someone's future has already been studied then the researcher can attempt to adjust for the differences in that variable. Matching and attempting to adjust for confounding variables is much more likely to yield unreliable results than performing an experiment because researchers can fail to address a lurking variable, misunderstand the interaction of variables when trying to control for confounding variables, or simply face a situation where a relevant variable is not easily quantifiable, but it is often the only option that researchers have when faced with ethical, practical or financial limitations. Give another example where, for practical or ethical reasons, an experiment could not be performed and researchers would need to rely on observational data.

If the research singled out based off of race/gender, which some people could find offensive

Class Activity 3.5: Data Ethics

While experiments tend to create more challenging ethical questions because treatments are being imposed on subjects, both observational studies and experiments have forced us to ask what qualifies as ethical behavior when collecting data. Let's look at three controversial studies from the 20th century to begin our discussion of the progression of data ethics: The Tuskegee Study, The Milgram Experiment, and the Stanford Prison Experiment.

Class Activity 3.6: The Tuskegee Study

In 1926, syphilis was recognized as a major health problem in the United States with a prevalence of 35% among the reproductive age population. In 1932, the Public Health Service partnered with the Tuskegee Institute to study the progression of syphilis. 600 poor black sharecroppers were recruited, 399 with syphilis and 201 without, and they were given free medical exams and meals as well as burial insurance in exchange for being treated for what they were told was "bad blood." The study was supposed to last less than a year but continued until 1972. In the early 1940s, penicillin became available and was accepted by the medical community as the treatment of choice for syphilis. Despite the existence of an effective treatment, the researchers prevented any treatment for the participants in the Tuskegee study. The study ended in 1972 when the Associated Press ran a story about the study and public outcry led to an advisory panel reviewing the study, deeming it unethical, and ending the study. In 1973, a class-action lawsuit was brought on behalf of the study participants and their families. In 1974, the case was settled out of court for 10 million dollars and the National Research Act was signed into law and regulations were passed that controlled how studies that gather data from human subjects are conducted.[14]

To this day, data ethics is a hotly debated topic, but a few basic standards have been accepted. Individuals in a study must give **voluntary informed consent**, individual data should be kept **confidential**, and studies must be approved in advance by an **institutional review board**. Voluntary informed consent means that subjects must be made aware of the risks of the study and then choose to participate and give consent in writing. Information being kept confidential does not mean that studies are anonymous, where names of the subjects are not known to anyone including the study director. It means that personal information will not be released except for in the form of a summary of data from a large group of subjects. Institutional review boards determine whether proposed studies protect the rights of the human subjects who would be participating in the study.

Class Activity 3.7: The Milgram Experiment

While the Tuskegee study was being performed in Macon County, Alabama, a social psychologist named Stanley Milgram was performing a study on the nature of obedience at Yale University. Milgram said he was inspired to do the study after observing the obedient behavior of Germans who were persuaded by Nazis to participate in the slaughter of innocent people. To study the nature of obedience, Milgram ran his experiment from 1961 to 1962 to see how far humans will go when encouraged by an authority figure.

In the Milgram Experiment, the subject arrives and is told that he or she will be participating in a study on memory. It appears that the subject has been randomly assigned to be the teacher and that another volunteer has been randomly assigned to the learner. It is explained that the teacher will be administering electric shocks of increasing intensity to the learner who is in another room to see what effect the shocks have on the learner's ability to successfully answer a memory question. In reality, it is all a ruse. The "learner" is in fact part of the study who is not receiving shocks at all but instead the teacher is hearing prerecorded responses that he or she thinks is from the person the teacher met upon arrival. The study is not of memory at all but is trying to see how high up the shock generator the teacher will go as voltage gets higher and higher and the prerecorded responses get more and more desperate for the teacher to stop. Stanley Milgram found that encouragement from the fake researcher in the room was enough to get 65% normal residents of New Haven to give seemingly harmful shocks up to 450 volts to a learner who earlier mentioned a heart condition and continually pleads for the teacher to stop.[15]

The Milgram Experiment was groundbreaking in terms of our understanding of human behavior and normally harmless people's ability to follow cruel orders when responsibility can be passed off on an authority figure. In addition, the shocks were not real and after the study the subject was informed about the true nature of the study and was able to reconcile with the "victim." Nonetheless, this experiment which was acceptable in the early 1960s would never be allowed today. What possible reasons could an institutional review board give to reject this study?

_____
_____
_____
_____

Class Activity 3.8: The Stanford Prison Experiment

In the summer of 1971, Stanford psychology professor Philip Zimbardo conducted a study to look at the relationship between guards and prisoners. 24 students were selected out of 75 who responded to an advertisement looking for male college students to participate in a psychology experiment. A coin was used to randomly assign half of the volunteers to be guards and the other half to be prisoners in a fake prison set up in the basement of the Stanford Psychology Department building.

Despite the fact that nothing more than a coin toss determined whether a volunteer would be a guard or a prisoner, the treatment of the prisoners by the guards during the study became increasingly abusive. Zimbardo served as superintendent of the study and for a time, permitted the abuse, but at the urging of psychology Ph.D. student Christina Maslach who was Zimbardo's girlfriend at the time and future wife, the study was stopped after six days.[16]

While the experiment yielded insight into the psychology of imprisonment, it also highlighted the importance of oversight in experimentation. After institutional review boards approve studies, they continue to monitor the progress of the study to ensure that subjects' rights and welfare continue to be protected.

-1 & 1 = high correlation
0 = no correlation

Homework Assignment 3.4: An Ethical Question

There has been a lot of progress in terms of protecting the rights of people involved in experiments, but there are many questions where consensus has not yet been reached. Choose one of the three following questions and justify an answer below.

1) If the welfare and rights of those involved in experiments must be valued above the interests of science and society, is it ever possible to justify the use of a placebo group? If it is, when is the use of a placebo group permissible? Does this change if subjects are being treated for something immediately life-threatening?

2) The law requires that clinical trials prove that new drugs work and are safe before they are approved for market. This is not the case for surgeries. To do a proper comparative experiment on surgeries, the control group's placebo would be a sham surgery. Without the randomized comparative experiment, it is difficult to conclude whether the surgery is effective or if we are merely witnessing the placebo effect. With the randomized comparative experiment, volunteers could be undergoing surgery with no actual treatment provided. When should surgeries be tested using randomized comparisons?

3) The door-in-the-face technique is often studied in introductory psychology classes. The idea is that people are more likely to agree to a request if they have already refused some larger request. The technique can be studied in two steps. The first step it to poll a group of people about their willingness to perform some task, e.g. filling out a 5 minute survey. The second step is to ask a different group of people if they will perform some burdensome task, e.g. watch a 3 hour video and, assuming they say no, ask if they would be willing to fill out the same 5 minute survey. The theory is that people will be more likely to fill out the 5 minute survey after "slamming the door" on the first request. The people being polled did not give voluntary informed consent to be a part of this experiment, however, if the nature of the study had been revealed, the experiment would not work. Is this okay? If it is, what are the conditions under which it is okay to lack voluntary informed consent?

_____
_____
_____
_____
_____
_____
_____
_____
_____

Part B: Analyzing Data

Chapter 4. Displaying and Describing Data

Now we can start to discuss how to work with the data that we produce. Let's look at some data produced by the Current Population Survey in Table 4.1 below. The Current Population Survey samples over 50,000 households each month to gather economic and social data on over 100 million U.S. households.[17]

Table 4.1 Level of Education of Males 25 to 64 Years Old in the United States, 2013

Level of Education	Number of persons (thousands)	Percent
Less than 9th Grade	3,414	4
9th to 12th Grade	5,852	7
High School Graduate or GED	24,665	31
Some College; No Degree	13,395	17
Associate Degree	7,771	10
Bachelor's Degree	16,646	21
Master's Degree	6,142	8
Professional Degree	1,357	2
Doctorate Degree	1,586	2
Total	80,829	100

Source; U.S. Census Bureau, Current Population Survey, 2013.

The last two columns in Table 4.1 give the **distribution** of the variable "level of education." A distribution of a variable explains how often a variable takes different values. In this table, the distribution is given both as a count, the number of persons in each group, and as a rate, the percent. The careful observer may also note that the sum of the counts is 80,828 instead of 80,829 as the table says. This is an example of the type of roundoff errors that can exist when inexact values are used after rounding.

Different tabular and graphical representations of distributions are appropriate for different situations. If you want to know how many males between 25 and 64 took some college courses but did not earn a degree, the count of approximately 13,395,000 will be more useful than the rate of 17 percent. If you want to know about what proportion of the male population between 25 and 64 has a doctorate, the rate of 2 percent will be more useful than the approximate count of 1,586,000. In this case, it would be possible to use a bar graph or pie chart to represent this data as seen in Figure 4.2 and Figure 4.3 respectively. If, however, we were comparing the percent of males with Master's and females with Master's, it would be appropriate to use a bar graph but not a pie chart. Pie charts need to show parts of a whole.

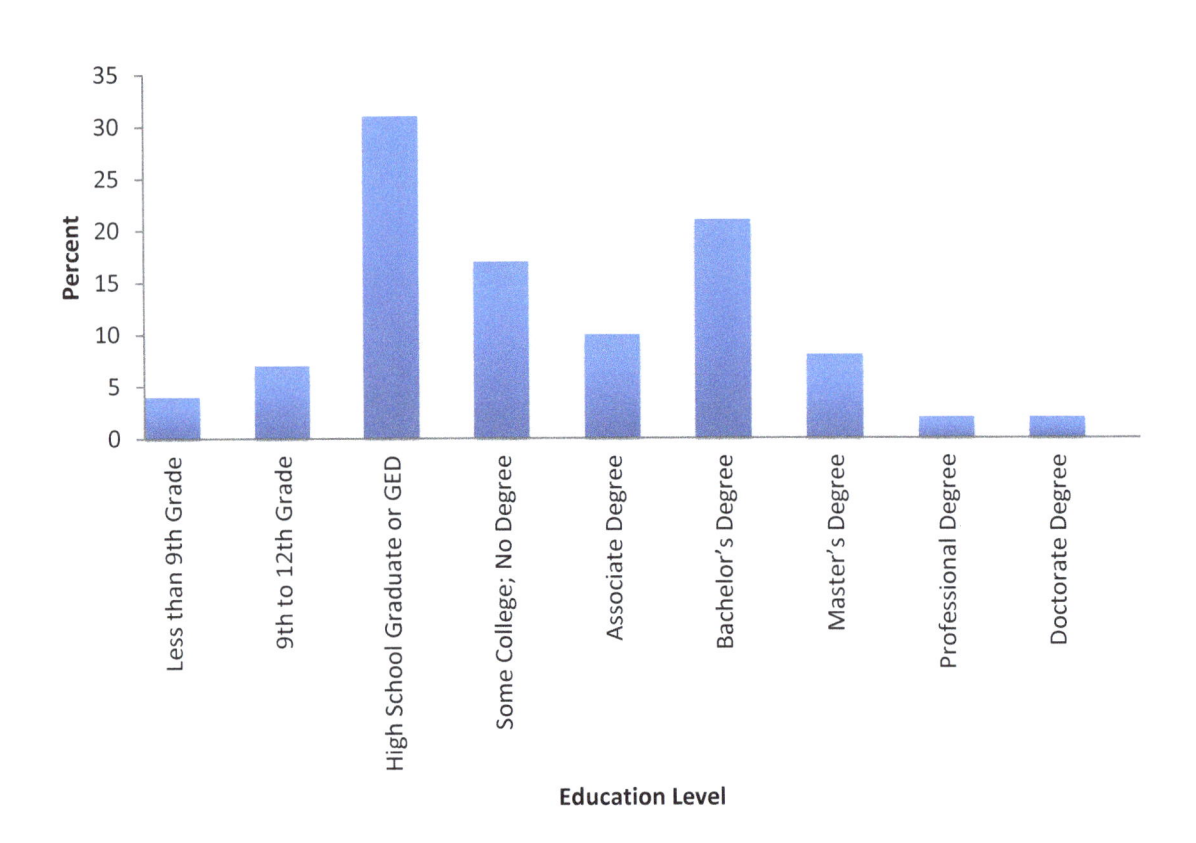

Figure 4.2 Bar graph of distribution of education among males aged 25 to 64 in the United States in 2013.

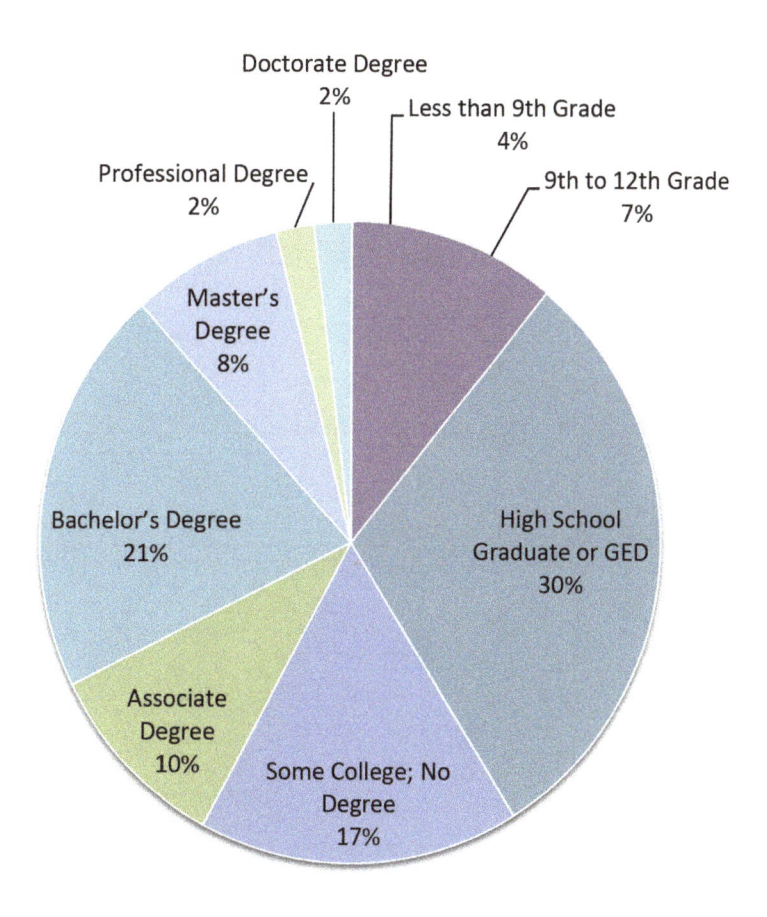

Figure 4.3 Pie chart of distribution of education among males aged 25 to 64 in the United States in 2013.

Bar graphs and pie charts had already been introduced in chapter 1. Now let's discuss three other methods for displaying quantitative variables, **dotplots**, **stemplots**, and **histograms**.

Dotplots are occasionally used for categorical data, but it is the simplest graph used to display quantitative data and is ideal for small data sets. In a dotplot, every individual is represented by a dot placed above the value that individual takes for the quantitative variable being considered.

A stemplot breaks observations into a stem and a leaf. The stem represents all but the final digit and the leaf represents the final digit. For example, for 293 the stem would be 29 and the leaf would be 3. The stemplot then organizes stems in a vertical column with the smallest on the top and then organizes the leaf values in order from least to greatest to the right of their corresponding stems. Stemplots are also useful with small data sets. If we are dealing large amounts of data, a histogram is a more common way to graph a quantitative variable.

A histogram is very similar to a bar graph with three key differences. 1, a bar graph can represent categorical data but a histogram always represents quantitative data. 2, each bin, the name for histogram bars, in a histogram must represent a range of values that is equal to the range of every other bin in the histogram and is inclusive on the lower end and exclusive on the upper end. That means that, in a histogram, the intervals 5.0≤x<6.0 can be next to 6.0≤x<7.0, but could not be next to 6.0≤x<8.0 or 6.0<x≤8.0. 3, in a histogram there is no space between the bins like you find in a bar graph.

Class Activity 4.1: Dotplots
How many siblings do you have?

Student #1		Student #9	
Student #2		Student #10	
Student #3		Student #11	
Student #4		Student #12	
Student #5		Student #13	
Student #6		Student #14	
Student #7		Student #15	
Student #8		Student #16	

Make a dotplot of the data.

0 1 2 3 4 5 6 7 8 9 10

Handwritten notes:
- Convenience Sampling } BIASED
- Voluntary Response Sampling
- ★ Polling Bathroom Activity
- ★ Telephone survey & in-location survey

Class Activity 4.2: Stemplots

How tall are you in inches? If you are 6'2", then you are 6x12 + 2 = 74 inches tall. Record the height in inches of every person in the class below rounded to the nearest inch.

Student #1		Student #9	
Student #2		Student #10	
Student #3		Student #11	
Student #4		Student #12	
Student #5		Student #13	
Student #6		Student #14	
Student #7		Student #15	
Student #8		Student #16	

Make a stemplot of these data.

```
4 |
5 |
6 |
7 |
8 |
```

↓

```
4 |
5 |
6 |
7 |
8 |
```

Variability = how spread out the values of the sample statistic are when we take many samples

Bias = consistent repeated deviation of from the sample statistic from the population in the same direction

to get random sampling
→ table of digits
→ random number generator

margin of error = $\frac{1}{\sqrt{n}}$

SRS
as long as each member has an = possibility of getting selected

flip a coin
shuffle a deck
RandInt
Pick a jury

Becomes statistically significant when n = 30

49

4 Bullseyes

Perameter = population
Sample = statistic

Low/High Bias vs High/Low Variability

Class Activity 4.3: Histograms

The following are the number of speeches in each of Shakespeare's 37 plays.[18] A speech is defined as any number of words spoken by a character, or a stage direction.

```
1,361  1,309  1,301  1,250  1,240  1,224  1,181  1,163  1,123
1,062  1,034  1,031  989    987    987    979    965    949
943    921    895    884    872    870    853    814    788
787    765    756    722    702    664    662    636    621
605
```

Make a histogram of the data. The x-axis will be "Number of Speeches" and the y-axis will be "Number of Plays." Start the x-axis at 500 and use a bin width of 125. The scale should run from 500 to 1,375. Be careful when determining which bin should include Hamlet (1,250 speeches).

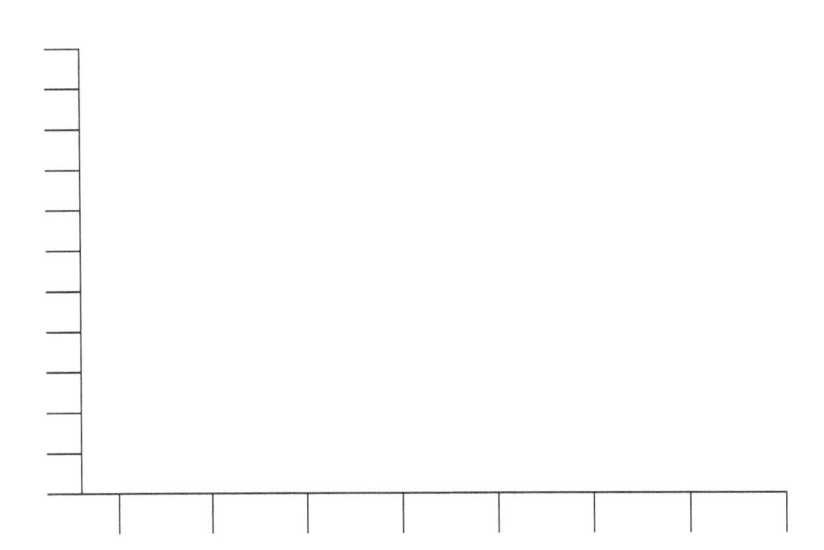

What would this, or any, histogram look like if the bin width were too small? For example, what if we used 5 for our bin width instead of 125?

_____

What would this, or any, histogram look like if the bin width were too large? For example, what if we used 1,000 for our bin width instead of 125?

_____

Homework Assignment 4.1: Displaying Distributions Graphically and Interpreting Histograms

There were 18 employees of MBA who taught at least one math class in the 2014-15 school year. The number of classes taught by those 18 teachers is shown below.

| 4 | 4 | 3 | 2 | 4 | 4 | 4 | 1 | 4 |
| 2 | 4 | 4 | 3 | 4 | 4 | 4 | 2 | 4 |

Make a dotplot of the data.

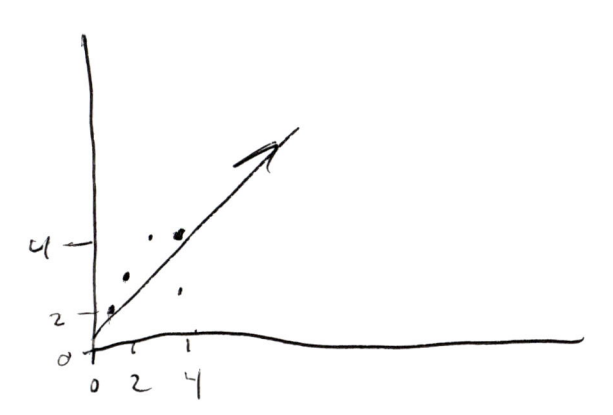

The following are the number of points that MBA's Varsity Basketball Team scored in the 27 games they played during the 2014-15 season.

57	76	38	32	59	69	62	57	77
63	65	50	56	67	59	59	43	41
46	58	43	52	36	58	38	49	47

Make a stemplot of the data.

```
3 | 8 2 6
4 | 3 1 6 3 9 7
5 | 7 9 7 0 6 9 9 8 2 8
6 | 9 2 3 5 7
7 | 6 7
```

The following are the salaries of the 19 basketball players who played for the Memphis Grizzlies at some point during the 2014-15 season.[19]

$16,500,000 $15,829,688 $9,200,000 $8,760,000
$5,450,000 $5,000,000 $3,911,981 $3,000,000
$2,077,000 $1,344,120 $967,500 $816,482
$816,482 $507,336 $507,336 $214,869
$86,504 $72,234 $10,694

Make a histogram of the data. The x-axis will be "Salaries" and the y-axis will be "Number of Players." Start the x-axis at $0 and use a bin width of $3,000,000. Be careful when determining which bin should include Kosta Koufos ($3,000,000).

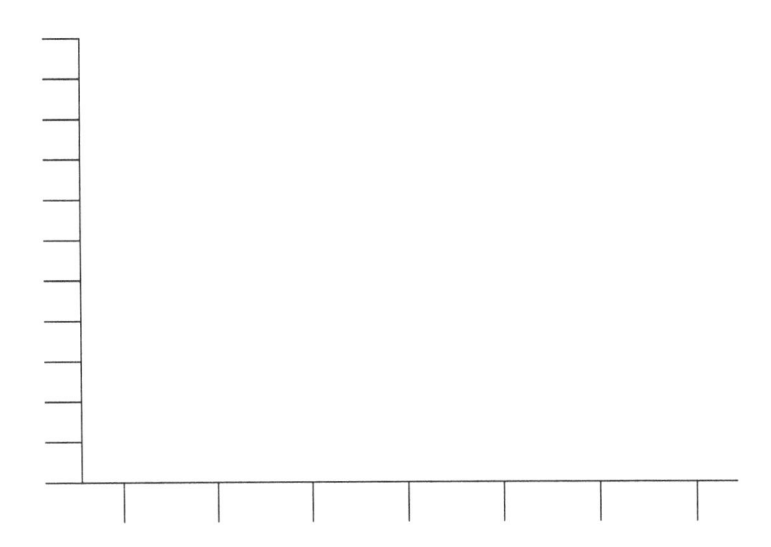

The following is a histogram of test scores recorded as integers. Which of the following statements is true?

I. At most 4 students scored a 100
II. If the passing score is 70, most students passed
III. At least 4 students scored above 90

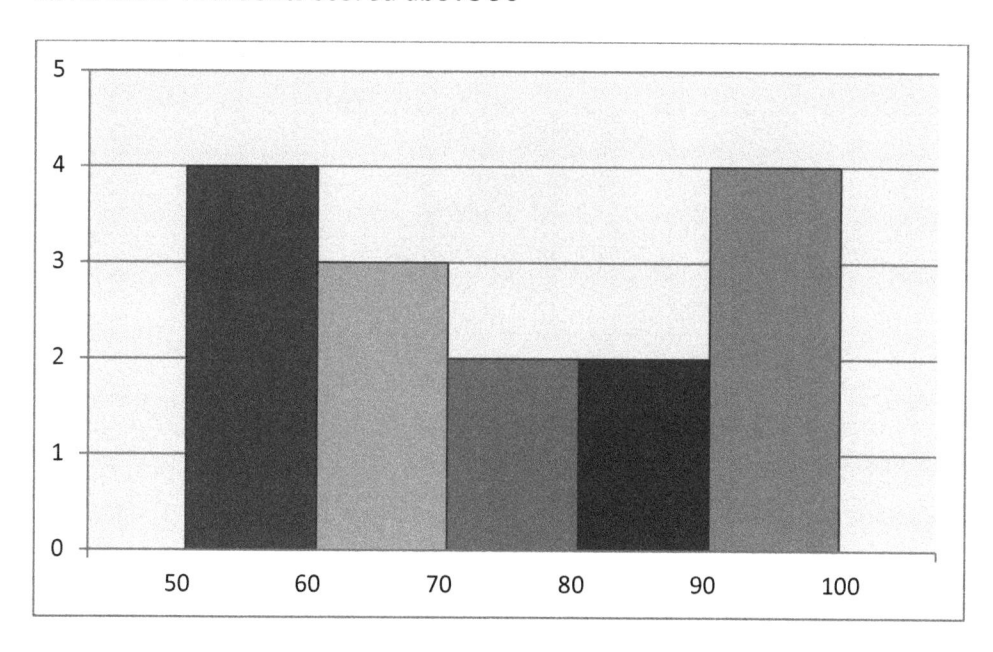

a) I only

b) II only

c) III only

d) I and II

e) II and III

f) I and III

g) I, II and III

Class Activity 4.4: Making a Histogram on Your Calculator

Now let's make the same histogram of Grizzlies salaries you did at home, but this time let's do it on your calculator.

On a TI-83 or TI-84, press the STAT button and then select 1:Edit... This is where you will enter your data (the Grizzlies salaries). If you there are values in any of the lists (L1, L2 etc.), use your arrow keys to select the list and hit the CLEAR button. Once the lists are clear, enter the salaries in L1. When you are done, your screen should look like the following:

Now hit 2ND and Y= to select the STAT PLOT option. Once you get to the STAT PLOT display, hit ENTER or 1 to select Plot1. Turn the plot On and select the top right graphing option so that your screen looks like the following:

If you hit ZOOM and then select 9: ZoomStat now, you should see a histogram, but it will not look like the histogram you created for homework. This is because you have not given the calculator instructions for where your bins should begin and what the widths of the bins should be. To do so, hit the WINDOW button and change the values so that your screen looks like the one below and matches the intervals used on the homework.

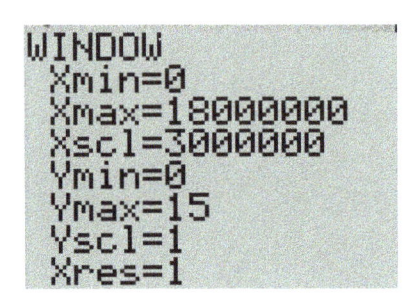

Now if you hit the GRAPH button, the histogram created should look like the one you created previously. If you want to jump from bin to bin and see how many individuals fall within that specified interval, push the TRACE button. When you hit the left and right arrow buttons, the calculator will display the interval selected and the number of individuals that fall in that interval. For example, if you move to the second bin, the calculator tells you that 4 Memphis Grizzlies made greater than or equal to $3,000,000 and less than $6,000,000 during the 2014-15 season. This calculator display is shown below.

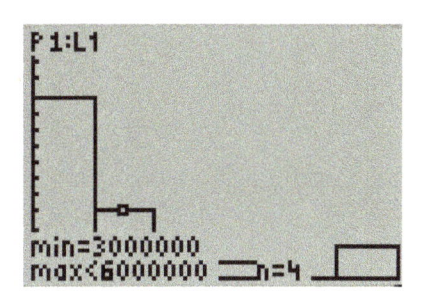

Class Activity 4.5: Beginning to Describe Distributions

An astute student will note that information is lost when we turn our data set into a histogram. For example, if you look at the list of salaries, you can see exactly how high the Grizzlies highest salary was during the 2014-15 season. It was $16.5 million (this salary was Zach Randolph's). If you look at the histogram, all you could say was that the highest salary was greater than or equal to $15 million and less than $18 million. The same issue is confronted when describing distributions with terminology or numbers. Information will almost always be lost when data are displayed graphically or described, so a statistician has to weigh the importance of the information being lost against the burdensome task of interpreting large amounts of data.

After creating a statistical graph, you should get in the habit of describing the overall pattern of the distribution. First, try to describe the shape of the distribution. A distribution with one clear peak is called **unimodal**. A distribution with two peaks is called **bimodal**. Any distribution with two or more peaks is called **multimodal**. The **mode**, often written as $\tilde{x}$ (x-tilde), is the observation that occurs most often in a data set. Strictly speaking, unimodal means that there is a single mode which is where the one peak occurs and bimodal means that there are two numbers that occur with the greatest frequency. In practice, however, if a distribution has two peaks or local maxima, even if one peak is higher because there is only one value that occurs with the greatest frequency, that distribution will often be referred to as bimodal.

The **median**, or center, is the middle value if the observations were listed smallest to largest. If the graph of the observations to the left of the median approximately mirrors the values to the right of the median, the distribution is said to be **symmetric**. If the observations to the right of the median extend much farther out than the observations to the left of the median, the distribution is said to be **skewed right**. If the observations to the left of the median extend much farther out than the observations to the right of the median, the distribution is said to be **skewed left**. It should be noted that while we will simply refer to distributions as approximately symmetric, skewed right or skewed left, skewness can be quantified. When it is quantified, approximately symmetric distributions have skewness close to zero. The more skewed right a distribution is, the more positive the value is that describes its skewness. The more skewed left a distribution is, the more negative the value is that describes its skewness.

The **range** is the difference between the largest and smallest values and is one way of evaluating the variability or spread of a distribution. At times, you may see a range given excluding **outliers**. Outliers are observations that deviate significantly from the overall pattern in the distribution. That is a lot of new terms. Let's see if we understand them by trying to describe the distribution of Memphis Grizzlies salaries that we graphed by hand and on our calculator previously.

Strictly speaking, is the distribution of Memphis Grizzlies salaries unimodal or bimodal? _____

Simply looking at the histogram of the salaries, would you describe the salaries as unimodal or bimodal? _____

What is the median salary for the 2014-15 Memphis Grizzlies? _____

Would you describe the distribution as approximately symmetric, skewed right, or skewed left? _____

What is the range of the salaries? _____

Do you think that any of the salaries qualify as outliers? If so, which?

_____

_____

Homework Assignment 4.2: Describing and Graphing Batting Averages

The following are the batting averages of the fifteen 2014 Atlanta Braves who had at least 100 at bats[20]:

.288 .263 .271 .270 .244 .208 .263 .251 .197
.204 .245 .162 .266 .212 .248

Make a histogram of the data by hand or on a calculator and transfer the result to the paper. The x-axis will be "Batting Average" and the y-axis will be "Number of Players." Have the x-axis start at .150 and end at .300 with bin widths of .025.

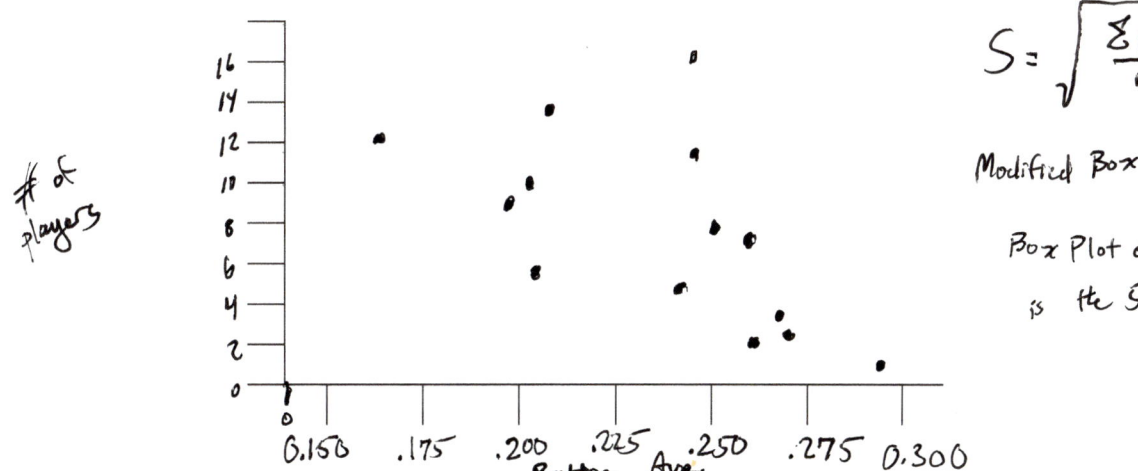

$$S = \sqrt{\frac{\Sigma(x-\bar{x})^2}{n-1}}$$

Modified Box Plot = includes outliers

Box Plot essentially is the 5-Number summary

Would you describe the batting averages as unimodal or bimodal?

__bimodal__

What is the median batting average for 2014 Braves with 100+ at bats?

__0.244__

Would you describe the distribution as approximately symmetric, skewed right, or skewed left?

__skewed left__

What is the range of the batting averages?

__0.126__

Do you think that any of the batting averages would qualify as an outlier? If so, which?

__Yes; 0.162__

How do you think the range would be affected if we looked at all 2014 Braves who had at least 1 at bat, instead of requiring at least 100 at bats?

The info would be inaccurate because of the smaller size in data

Tough one. Let's say we look batting average histograms for all teams in Major League Baseball where only players who have 40 or more at bats in a season are included. How do you think American League teams and National League teams would differ in terms of skewness? Note: In the American League a designated hitter is allowed to bat in place of the pitcher.

American League hitters would have on average a higher % because of the ability to handpick players for designated hitting.

Class Activity 4.6: Describing Distributions with Numbers

To learn more about describing distributions, let's do an activity. The teacher is going to tell you a food item and then pass out pieces of paper to every student. On your piece of paper, take your best guess at the number of calories in that food item, write your name, fold the piece of paper, and pass it forward. Today the data with which we are going to work are the guesses by the students in class. Record the guesses below.

Food: _____

Student #1		Student #9	
Student #2		Student #10	
Student #3		Student #11	
Student #4		Student #12	
Student #5		Student #13	
Student #6		Student #14	
Student #7		Student #15	
Student #8		Student #16	

Before we reveal the number of calories, let's analyze this data.

First, let's find the median. As we said before, the median is the middle number if the observations are ordered from smallest to largest. You may also hear it referred to as the second quartile (Q_2). If there is an even number of observations and thus two middle-most numbers, add those two numbers together and divide by 2 to get the median.

Median: _____

Now let's find the **first quartile** (Q_1) and **third quartile** (Q_3). The first quartile is the median of just the observations which are less than the overall median. The third quartile is the median of just the observations which are more than the overall median. If there is an odd number of observations in the entire data set, do not include the overall median when finding Q_1 and Q_3. If there is an even number of observations, the two middle-most values are included when finding Q_1 and Q_3. The first middle-most value is used when finding Q_1 and the second middle-most value is used when finding Q_3.

Q_1: _____ Q_3: _____

You may know the first quartile and third quartile by different names, the 25th and 75th percentile respectively. Colleges report the 25th and 75th percentile of the SAT and ACT scores of its incoming class. This can be useful in determining if a college is likely to grant you admission as the 25th and 75th percentile in this context essentially tells you the range of the scores of the middle 50% of the class.

Finally, let us find the **minimum** and **maximum** values. These are the smallest and largest observations in the data set respectively.

Minimum: _____   Maximum: _____

These five values in the following order make up the **five-number summary** of a distribution:

Minimum Q_1 Median Q_3 Maximum

Up to this point, we have simply defined an outlier as an observation that deviates significantly from the overall pattern in the distribution. This is a subjective definition. Now we will introduce the most common rule used to objectively define outliers. An observation is an outlier if it is more than 1.5 times the **interquartile range** (IQR) above Q_3 or below Q_1. The interquartile range is the difference between the first and third quartiles, i.e. $Q_3 - Q_1$. That is:

An observation is an outlier if it is

1) greater than Q_3 + 1.5(IQR) or

2) less than Q_1 − 1.5(IQR)

Let's find the IQR and see if there are any outliers.

IQR: _____   Outlier(s): _____

The five-number summary of a distribution can stand on its own, or it can be used to create a **boxplot**, which is also known as a box and whisker plot. A boxplot is drawn above a labeled and numbered horizontal axis. A box or rectangle is drawn that starts at Q_1 and ends at Q_3. A vertical line is drawn down the middle of the box at the median and lines (or whiskers) extend from the sides of the boxes to the smallest and largest values that are not outliers. Outliers are represented as separate points.

Draw a boxplot of our data below.

Now let's reveal the actual number of calories and see which guess was closest.

Actual Number of Calories: _____

Homework Assignment 4.3: Making an MBA Football Boxplot

The following are the number of points scored by MBA in each of the 13 varsity football games played during the 2014 season:

26 42 20 49 52 42 44 21 42
58 43 31 10

Find the five-number summary of the data: 10, 23.5, 42, 46.5, 58

Are any of these values outliers? __No__

Draw a boxplot of the data below.

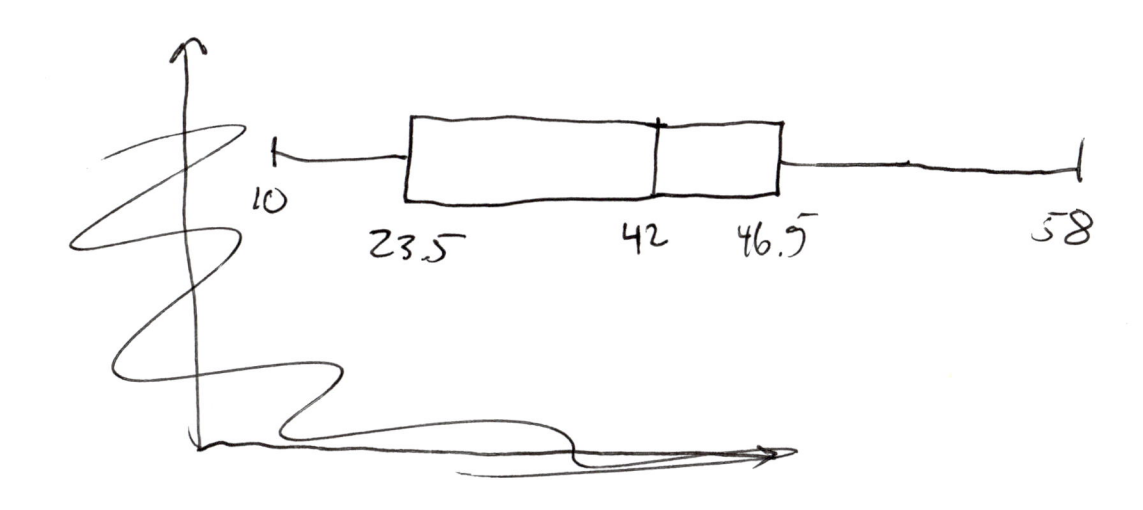

Class Activity 4.7: Making a Boxplot on Your Calculator

Now let's make the same boxplot of points scored by MBA's offense that you did at home, but this time let's do it on your calculator.

Press the STAT button, select 1:Edit…, and enter the data in L1. Push 2ND and Y= to select the STAT PLOT. Select Plot1, turn the plot On and select the bottom left graphing option so that your screen looks like the following:

If you hit ZOOM and then select 9: ZoomStat now, you should see our boxplot. If you hit the TRACE button, you will be able to jump between the numbers in our five-number summary by hitting the left and right arrow keys. Below you can see that the calculator shows that Q_3 = 46.5 when you hit TRACE and jump to the right side of the box.

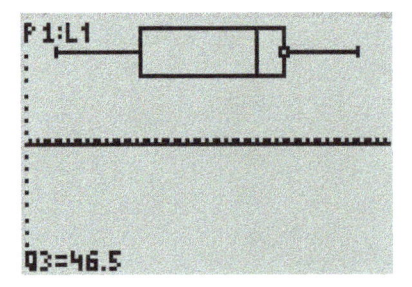

In this particular boxplot, there were no outliers. When there are, they are graphed by the calculator as individual points and you can jump from outlier to outlier using the TRACE function.

Class Activity 4.8: Mean and Standard Deviation

One advantage of the five-number summary and the boxplot is that it can quickly show you the skew of a distribution. A distribution can also be described by its **mean** ($\bar{x}$) and **standard deviation** (s). Mean and standard deviation are used more frequently than five-number summaries and are especially useful when working with distributions that are roughly symmetric. The mean, median and mode are known as measures of central tendency and each have their advantages and disadvantages in terms of describing a distribution. The mean, or average, of a distribution is found by taking the sum of all observations and then dividing that sum by the number of observations.

What if in the 2014 football game against McCallie MBA scored 100 points instead of scoring 58? How would that affect the median that we found in Homework Assignment 4.3? _____

What about the mean? How would the mean change if MBA scored 100 points instead of scoring 58 points in that game? _____

As you can see, outliers can have a significant effect on mean. They do not have as great an influence on the median. As a result, we say that the median is *resistant*.

It is not necessarily good or bad to be resistant but it is important to understand that if any value is changed in a distribution, it will affect the mean. This will not always be the case when finding median and mode.

Standard deviation is often defined as the average distance of observations from their mean. Strictly speaking, that is the definition of the mean absolute deviation (MAD) but both standard deviation and MAD help describe how spread out a distribution is. If all of the observations in a distribution are the same, both the MAD and standard deviation would be zero. As numbers become more spread out, both the MAD and standard deviation increase. While MAD has simplicity on its side, standard deviation is used much more often.

The following formula is for the standard deviation of a sample:

$$s = \sqrt{\frac{\Sigma(x-\bar{x})^2}{n-1}}$$

Like many formulas we will encounter in this course, it looks much more complicated than it actually is. Sample standard deviation can be found in four easy steps:

1) Subtract the mean of the sample from each observation to find the deviation from the mean.

2) Square each of these deviations and sum these squared deviations.

3) Divide this sum by one less than the number of observations, n – 1. This quotient is called the **variance**.

4) Take the square root of variance to find the standard deviation.

Let's practice finding sample standard deviation. Some MBA students are asked how many pets they have had. The following were the responses:

Number of Pets: 3   5   0   6   2   3   2

What is the sum of the observations? _____

What is the mean ($\bar{x}$) number of pets? _____

Find the difference between each student's number of pets and the mean (x - $\bar{x}$)?

Deviation: ___ ___ ___ ___ ___ ___ ___

Square these deviation values.

Squared deviation: ___ ___ ___ ___ ___ ___ ___

Find the sum of these squared deviation values. _____

Find n – 1, the number of observations less one. _____

Find the sum of the squared deviation values divided by n – 1. This is the variance. _____

Take the square root of variance to find standard deviation. _____

Normally, finding standard deviation is challenging to do without the aid of a calculator or computer. Let's find out how many pets everyone in class has had in his life. For homework, you will find standard deviation using Microsoft® Excel.

Student #1		Student #9	
Student #2		Student #10	
Student #3		Student #11	
Student #4		Student #12	
Student #5		Student #13	
Student #6		Student #14	
Student #7		Student #15	
Student #8		Student #16	

Homework Assignment 4.4: Finding Standard Deviation with Excel

1) On your personal or a school computer, open Microsoft® Excel. Create a new workbook and save it as PetStandardDeviation.xlsx.

2) We will start off by entering the pet values in column B of the first worksheet. In cell B1, type "Pets" and in B2 and below, type the values gathered at the end of class. While your values will almost certainly be different and you will likely have more observations, your worksheet should look like the following when you have finished this step.

3) In column A, enter the following labels in the rows immediately following the last observation value: "Sum," "Count," "Mean," "Variance," and "Standard Deviation." You may widen column A so that all the labels fit in their respective cells.

4) Click the cell directly to the right of the "Sum" label, then click **Autosum** near the top of the screen. **Autosum** automatically enters the equal sign, the SUM function and the range of numerical values above. If **Autosum** does not work, you can click the cell and then type "=SUM(" and then select the range of cells above that contain our numerical data.

5) Click the cell directly to the right of the "Count" label, then type "=COUNT(" and then select the same range of cells containing our numerical data that we selected in step 4. The COUNT function counts the number of observations in the range. Again, you should have different values and likely a different number of observations, but your worksheet should now look roughly like the following.

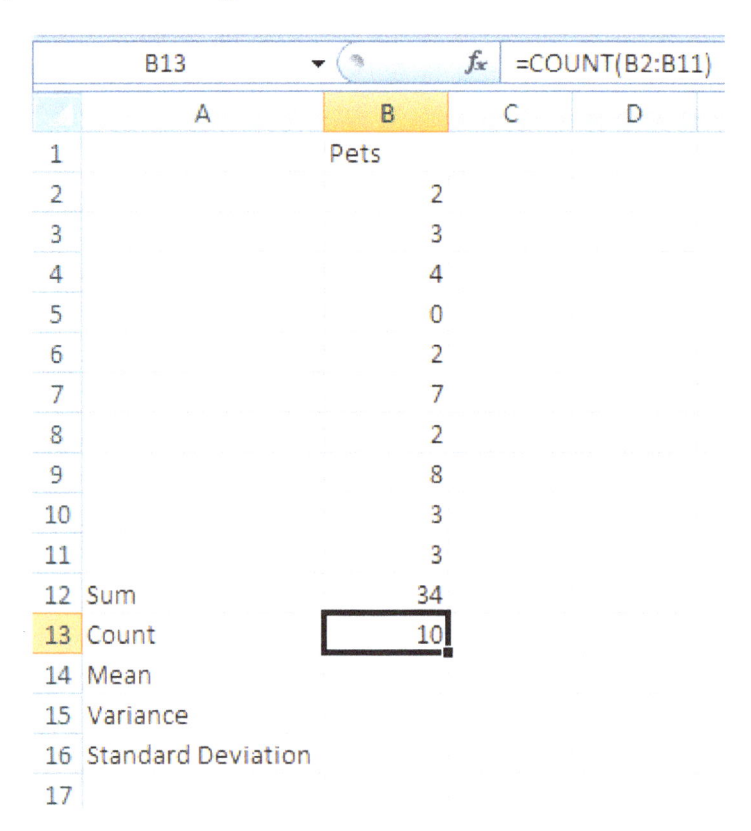

	A	B	C	D
			f_x	=COUNT(B2:B11)
	A	B	C	D
1		Pets		
2		2		
3		3		
4		4		
5		0		
6		2		
7		7		
8		2		
9		8		
10		3		
11		3		
12	Sum	34		
13	Count	10		
14	Mean			
15	Variance			
16	Standard Deviation			
17				

6) Click the cell directly to the right of the "Mean" label, then type "=" and then either select the cell to the right of the "Sum" label or type the name of that cell, then type "/" for division, and then select the cell to the right of "Count" or type the name of that cell.

7) Now we want to find out how much all of the observations deviate from the mean. Click on cell C1 and type "Deviation." In C2, type "=B2 –"and then select the cell that contains the mean value. On my Excel worksheet it is B14. I want my formula to read "=B2 - B14" instead of "=B2 - B14". You can type the $ symbol in manually or you can select the cell in the equation editor and then hit the F4 button to add these $ dollar signs. The dollar signs change the cell in the formula bar from a relative to an absolute reference. We are going to drag this box down to apply this formula to list all deviation values in column C. If the $ was not added to make the reference absolute and the box was dragged down, the formula in C3 would be "=B3 – B15" and the formula in C4 would be "=B4 – B16" instead of always referring to the mean value in cell B14. Now that we do have the correct formula in C2, highlight that cell and then drag from the bottom right corner of the cell down to the final cell to the right of an observation. This should calculate and display the deviation of every observation from the mean and your Excel sheet should look something like the following.

	A	B	C	D
1		Pets	Deviation	
2		2	-1.4	
3		3	-0.4	
4		4	0.6	
5		0	-3.4	
6		2	-1.4	
7		7	3.6	
8		2	-1.4	
9		8	4.6	
10		3	-0.4	
11		3	-0.4	
12	Sum	34		
13	Count	10		
14	Mean	3.4		
15	Variance			
16	Standard Deviation			
17				

C11 fx =B11-B14

8) Next, we will square all of these deviation values. Select D1, type "Deviation Squared" and adjust the width of the D column to make the label fit. One tip, if you want the column width to be just large enough to contain the characters contained in that column, double click on the right side of the column letter label box. For example, in this case, if you click the right of the cell containing "D" then the column will be made just wide enough to contain the "Deviation Squared" text. Activate D2 and type in the equation "=C2^2". The ^ is the exponentiation sign. Select D2 and drag from the bottom right down to the last row containing an observation value. Your Excel sheet should now look like the following except for the different values.

	A	B	C	D
1		Pets	Deviation	Deviation Squared
2		2	-1.4	1.96
3		3	-0.4	0.16
4		4	0.6	0.36
5		0	-3.4	11.56
6		2	-1.4	1.96
7		7	3.6	12.96
8		2	-1.4	1.96
9		8	4.6	21.16
10		3	-0.4	0.16
11		3	-0.4	0.16
12	Sum		34	
13	Count		10	
14	Mean		3.4	
15	Variance			
16	Standard Deviation			

(D11 selected, formula =C11^2)

9) Now we'll find the variance of this sample. To find sample variance, we sum the squared deviations and then divide by the number of observations less one. In the example spreadsheet, the formula is "=SUM(D2:D11)/(B13-1)" to add up all the deviation values and then divide by the one less than the count.

10) Lastly, to find sample standard deviation we will take the square root of variance. You can either raise variance to the ½ power or you can use the square root function in Excel. To use the square root function, type "=SQRT(" and then select the variance value. The value that Excel calculates is the sample standard deviation.

11) This exercise helps show how sample standard deviation is found while having Excel do the calculations for us. To check our answer, we can use the Excel function STDEV which will find sample standard deviation for us in one step. Select the cell to the right of the sample standard deviation value that we have found, type "=STDEV(" and then select the cells that contain our observations. If you have done this exercise correctly, the two cells to the right of Standard Deviation should be the same and your final worksheet should look like the following with different values.

	C16			f_x =STDEV(B2:B11)	
		A	B	C	D
1			Pets	Deviation	Deviation Squared
2			2	-1.4	1.96
3			3	-0.4	0.16
4			4	0.6	0.36
5			0	-3.4	11.56
6			2	-1.4	1.96
7			7	3.6	12.96
8			2	-1.4	1.96
9			8	4.6	21.16
10			3	-0.4	0.16
11			3	-0.4	0.16
12	Sum		34		
13	Count		10		
14	Mean		3.4		
15	Variance		5.822222		
16	Standard Deviation		2.412928	2.412928	

12) Save your document and email your teacher the Excel workbook as an attachment. Record the sample standard deviation that you found during the exercise below.

Sample Standard Deviation (s) = _____

Class Activity 4.9: Finding s on Your Calculator

You can also find sample standard deviation (s) on your calculator. Press the STAT button, select 1:Edit..., and enter the pet data in L1. Press the STAT button again, click the right arrow for CALC, select 1:1-Var Stats and then hit ENTER after "1-Var Stats" is displayed on your home screen. Note: By default, the calculator finds statistics on L1. If you want to find statistics on L2, L3, or any other list, simply type the list after 1-Var Stats before hitting enter to find statistics on that list.

When you hit ENTER, a list of information about L1 appears below. It should look something like the following.

```
1-Var Stats
 x̄=3.4
 Σx=34
 Σx²=168
 Sx=2.412928143
 σx=2.289104628
↓n=10
```

"Sx" represents the sample standard deviation and should be the same value that you found on your Excel worksheet. If you push the down arrow, you can see even more information, including the 5-number summary for this data set.

Class Activity 4.10: Standard Deviation for Samples and Populations

After exploring sample standard deviation, there are two questions that students often have. 1, "Why do we divide by n – 1 instead of n when finding s?" 2, "What is the similar value right below Sx in my 1-Var Stats?" The answers to these two questions are related.

When we asked the class about lifetime pet ownership, we were treating the class as a sample of a larger population, e.g. the population of all current MBA students. When we find the standard deviation of a sample (s), we are actually finding an estimate of the population standard deviation based on the sample. The notation for population standard deviation is the lower-case Greek letter sigma, σ. The formula for population standard deviation is very similar to our formula for sample standard deviation. The only difference is that after we sum the squares of the deviations, we divide by n instead of n – 1. This n – 1 will come up in future chapters. It is referred to as the degrees of freedom. If you treat the class pet data as all the data for our population, we would report the population standard deviation instead of the sample standard deviation. In that case, you would divide by n and you could check your result with 1-Var Stats by comparing your result to the σx value.

An insistent student would continue. "Okay, the lower-case sigma value represents population standard deviation, but why do we divide by n – 1 to find sample standard deviation?" The reason is this: by dividing by n – 1, the sample variance (s^2) becomes an unbiased estimate of population variance (σ^2). This means that if you looked at the sample variances of all the possible samples that could be drawn from the population, the average of those sample variances would equal the population variance! The proof of this is somewhat rigorous, but it can be confirmed by creating a small population, finding its variance, and then finding the variance of all the possible samples that could be chosen from that population. The average of those possible sample variances will equal the population variance.

Class Activity 4.11: Misrepresentation of Data

In this chapter, we have learned a great deal about how data can be represented graphically and numerically to help provide a helpful summary of information. While this process can be very useful, it can also end with misleading results if not done carefully. In this section, we will discuss some of the common ways in which data can be misrepresented.

To explain our first type of misrepresentation, we will introduce our final method for displaying quantitative data, the **line graph**. A line graph plots observations against the time at which they were measured and connects those observations with lines. Nashville natives know the city's population is growing, but how quickly? The following table shows the population of Davidson County according to the last five censuses.[21]

Year	Population
1970	448,003
1980	477,811
1990	510,784
2000	569,891
2010	626,681

The two line graphs on the next page both show this data graphically.

Nashville's Population: 1970-2010

[Line graph with y-axis from 0 to 700,000, showing population values approximately: 1970: 450,000; 1980: 480,000; 1990: 510,000; 2000: 575,000; 2010: 630,000]

Nashville's Population: 1970-2010

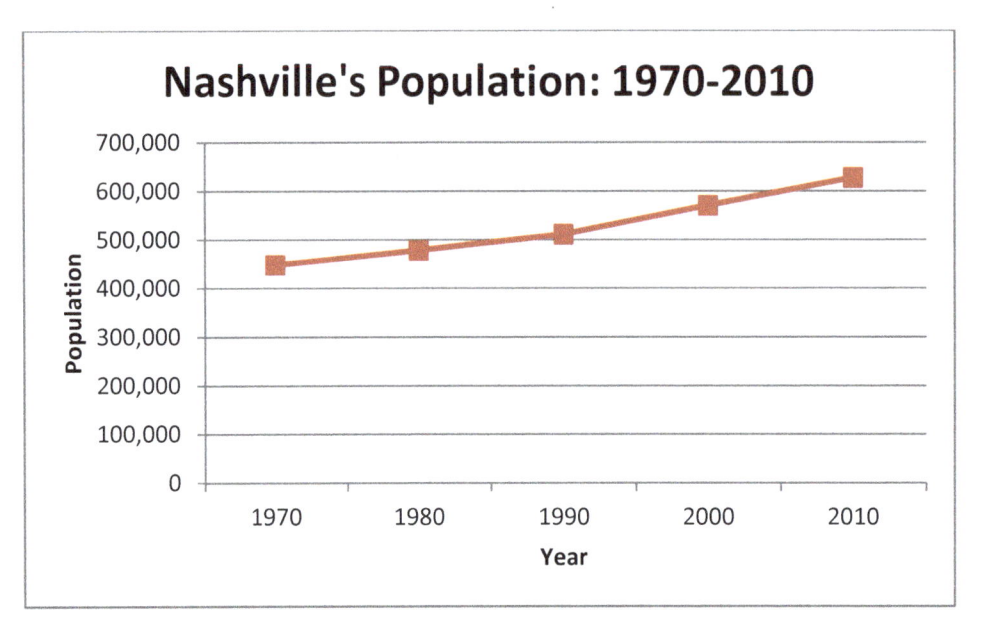

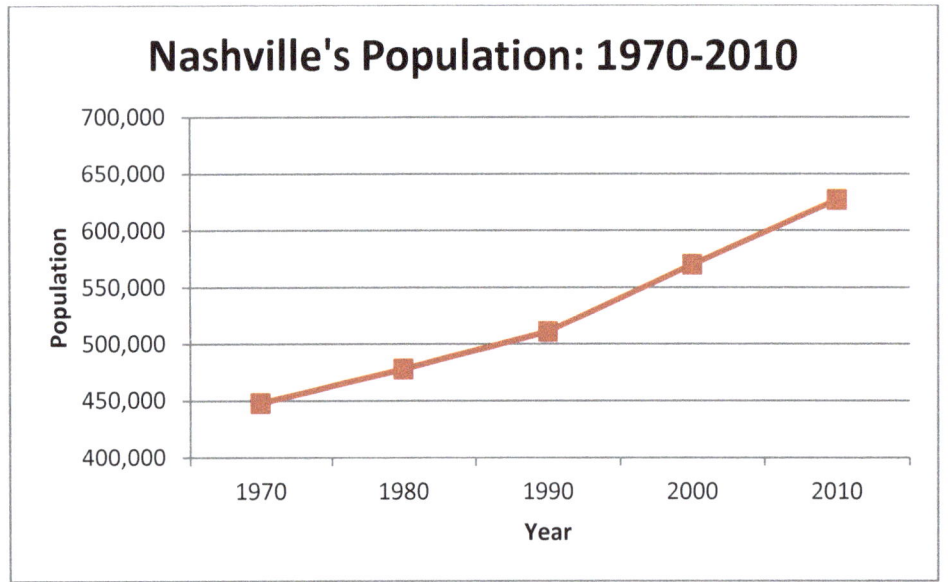

[Line graph with y-axis from 400,000 to 700,000, showing population values approximately: 1970: 450,000; 1980: 480,000; 1990: 510,000; 2000: 570,000; 2010: 625,000]

    Neither of these line graphs is strictly speaking incorrect, but the second graph is misleading. By starting the y-axis at 400,000 instead of 0, the upward **trend**, the long-term movement over time, appears to be much more dramatic. Unless you examine the scales carefully, you may think that Nashville's population has more than quadrupled over the past 40 years.

In which situations is it misleading to start a y-axis at a number other than zero? When would it be acceptable to start a y-axis at a number other than zero?

In many situations, data are misleading because information is left out. It seems that over the last couple decades, every couple years a movie becomes one of the top grossing films at the box office in history. In fact, as of July 2015, the top 10 domestic grossing films of all time were as follows.[22]

1. Avatar (2009)
2. Titanic (1997)
3. Marvel's The Avengers (2012)
4. Jurassic World (2015)
5. The Dark Knight (2008)
6. Star Wars: Episode I – The Phantom Menace (1999)
7. Star Wars: Episode IV – A New Hope (1977)
8. Avengers: Age of Ultron (2015)
9. The Dark Knight Rises (2012)
10. Shrek 2 (2004)

In Charles Wheelan's "Naked Statistics" he asks a fair question when considering the highest grossing films of all time: "Was [Shrek 2] really a greater commercial success than *Gone with the Wind? The Godfather? Jaws?*" The fact that only two films from the top 10 came from the 20th century should be a little alarming if we are using a film's domestic gross as the sole indicator of commercial success. Why? There are more people going to more theaters and paying more for tickets than in decades past. If you adjust for just this third issue, inflation, the list of the top 10 domestic grossing films of all time looks very different. The following is the top 10 list adjusted for ticket price inflation.[23]

1. Gone with the Wind (1939)
2. Star Wars: Episode IV – A New Hope (1977)
3. The Sound of Music (1965)
4. E.T.: The Extra-Terrestrial (1982)
5. Titanic (1997)
6. The Ten Commandments (1956)
7. Jaws (1975)
8. Doctor Zhivago (1965)
9. The Exorcist (1973)
10. Snow White and the Seven Dwarfs (1937)

One could argue that this list still fails to accurately assess the most commercially successful films in history. What other method could be used to rank films?

One popular yet notoriously misleading graphical representation is a **pictoral graph**, also known as a pictogram or pictograph. A pictoral graph is a bar graph that replaces bars with pictures. The visual appeal is clear. The problem is as follows. On a bar graph, when one value is twice another it is represented by a bar with twice the height, but the same size base. In a pictoral graph, when one value is twice another, one normally doubles the height *and* width so as to avoid stretching the picture. This results in a picture that is four times the size. As Darrell Huff explains in his *How to Lie with Statistics*, "the problem gets much worse when the pictures are represented as being three dimensional. Since these are pictures of objects having in reality three dimensions, the second must also be twice as thick as the first…Two times two times two is eight…And that indeed is the impression my ingenious little chart gives. While saying 'twice,' I have left the lasting impression of an overwhelming eight-to-one ratio."[24]

Homework Assignment 4.5: Lies, Damned Lies and Line Graphs

Open a web browser and go to www.google.com/trends. Google Trends shows how often a term has been searched over time on a line graph since 2004.

1) Find a term that shows an upward trend since 2004. _____

2) Find a term that shows a downward trend since 2004. _____

A line graph is said to show **seasonal variation** if there is a regular pattern in the graph that repeats annually.

3) Find a term that shows seasonal variation since 2004. _____

Three common ways in which representations of data can be misleading were discussed in this chapter: 1, a scale can start at a nonzero number to magnify or diminish the magnitude of a trend on a line graph 2, important information can be withheld when reporting data that, if included, would help answer the questions that the data is supposed to help answer and 3, a pictogram can magnify changes in values by adding a second or third dimension to the change.

4) Think of another way in which a representation of data (visual or numerical) can be misleading and give an example of the misleading method in practice.

_____
_____

Chapter 5. Studying Individuals within a Distribution

Class Activity 5.1: Finding Percentile

In this section we will introduce our first number used to describe an individual's location within a distribution: **percentile**. There are two commonly used definitions for percentile. The first way to define percentile is to say that the pth percentile is the value with p percent of observations *below it*. The second definition is that the pth percentile is the value with p percent of observations *equal to or below it*. For the purposes of this course, we will be using the second definition. You will see that which definition you use can make a significant difference when you are dealing with discrete variables and a small number of individuals but will rarely make a large difference when dealing with continuous variables and a large number of individuals.

To better understand percentiles, we are going to start by collecting some data from the class. In the table below, record the number of colleges to which every member of the class has applied and when you are done, make a dotplot of the distribution below the table.

Student #1		Student #9	
Student #2		Student #10	
Student #3		Student #11	
Student #4		Student #12	
Student #5		Student #13	
Student #6		Student #14	
Student #7		Student #15	
Student #8		Student #16	

```
 0    2    4    6    8   10   12   14   16   18   20
```

Now, let's calculate Student #8's percentile. To calculate the percentile, count the number of students who applied to the same number of schools or fewer (including Student #8), divide by the total number of students, and multiply that quotient by 100.

Student #8's number of college applications puts him at the _____ percentile in the class's college application distribution.

Class Activity 5.2: Using Standardized Tests to Understand Z-Scores

In this section we will introduce our second number used to describe an individual's location within a distribution: **z-score**. A z-score, often referred to as a standardized value, tells us how many standard deviations above or below the mean an observation falls. The following is the formula used to find an individual's z-score:

$$z = \frac{x - mean}{standard\ deviation}$$

To understand how this formula is used, write down an ACT score and an SAT score (on the 2400 scale) that you think are equally strong.

ACT score: _____   SAT score: _____

The mean ACT is approximately 21 and standard deviation is approximately 5. The mean SAT is approximately 1500 and standard deviation is approximately 300. Use those values and our new formula to find the z-scores of your two test scores.

ACT z-score: _____   SAT z-score: _____

A z-score of 1 means that the observation is one standard deviation above the mean. A z-score of -2.4 means that the observation is 2.4 standard deviations below the mean. A z-score of 0 means that the observation is equal to the mean. Explain below what the two z-scores mean regarding the position of those scores in the distribution of ACT and SAT scores and which score would be more impressive.

_____

Class Activity 5.3: Collecting Data about Theater Attendance

We are going to start by collecting some data from the class that will be used for the homework tonight. In the table below, record the total number of theater productions members of the class have attended at MBA (junior school play, collaboration with Harpeth Hall, one-acts, etc.). Only count a production once even if you went to multiple performances.

Student #1		Student #9	
Student #2		Student #10	
Student #3		Student #11	
Student #4		Student #12	
Student #5		Student #13	
Student #6		Student #14	
Student #7		Student #15	
Student #8		Student #16	

As a class, let's find the mean and standard deviation of this data set using 1-Var Stats on our calculator. Note: Since this data set represents our entire population of interest, we will use the population standard deviation (σ) instead of the sample standard deviation (s).

Total number of individuals =

Mean =

Standard deviation =

Homework Assignment 5.1: Interpreting Z-Scores and Percentiles

Find the percentile corresponding to the number of theater productions you have attended.
_____

Explain what this value means. _____

_____

Find the z-score corresponding to the number of theater productions you have attended.
_____

Explain what this value means. _____

_____

What would your percentile be if you used the other formula for calculating percentile, i.e. only counted the number of observations *below* instead of *equal to or below*? _____

Do you think your percentile or z-score more accurately describes where you fall in this distribution? Why? _____

_____

_____

Class Activity 5.4: Exploring Density Curves

If I want to display a distribution graphically, the appropriate type of graph often depends on the number of observations. If I were looking at the heights of students in this class, it would be appropriate to use a dotplot or stemplot. If I were looking the heights of all current MBA students, a dotplot and stemplot would be tedious to create and challenging to interpret. It would be more appropriate to use a histogram. For example, let's say that the following histogram represents the height in inches of all current MBA students with bin width of 1 inch from 60 to 76 inches.

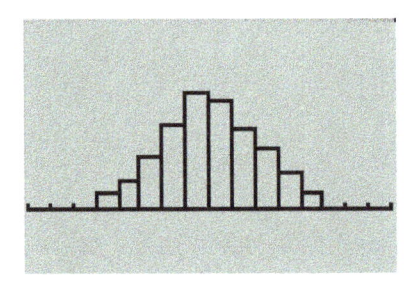

There is another option when considering distributions with a large number of observations and a regular pattern. It is called a **density curve**. A density curve is a continuous curve that captures the overall pattern of a distribution. For example, a density curve that generally captures the pattern of our height distribution might look something like the following.

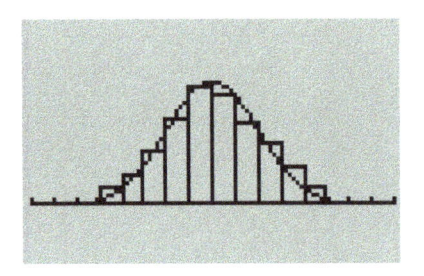

One key difference between histograms and density curves is that histograms tend to display counts, e.g. if there were 97 students greater than or equal to 73 inches but less than 74 inches tall, the height of that bin would normally be 97. A density curve, on the contrary, aims to attempt to capture the proportion of observations in any region under a curve, e.g. if 11% of students were students greater than or equal to 73 inches but less than 74 inches tall, the integral from 73 to 74 should equal .11. Look at that! Calculus strikes again! Since 100% of observations are supposed to be represented by the density curve, the area under any density curve is 100%, or 1.

The area in any region under a density curve represents the proportion of observations in the distribution that fall in that region. Since the median is the point at which half of the observations are above and half the observations are below, the median on the density curve is the equal-areas point, or the point at which half of the area is to the left and half of the area is to the right. If distributions are symmetric, the median and mean are found at the same location. Outliers in either direction, however, have a greater effect on the mean than the median. As a result, in distributions that are skewed left the mean is pulled to the left more than the median and in distributions that are skewed to the right, the mean is pulled more to the right than the median. If a density curve were made of a solid material, the mean would be the point at which the curve would balance.

Homework Assignment 5.2: Sketching Density Curves and Approximating Mean and Median

Below are three histograms. Sketch a density curve that describes the distribution well on top of the current histogram and then approximate the mean and median values.

1. A histogram of the distribution of results of 100 rolls of a fair die

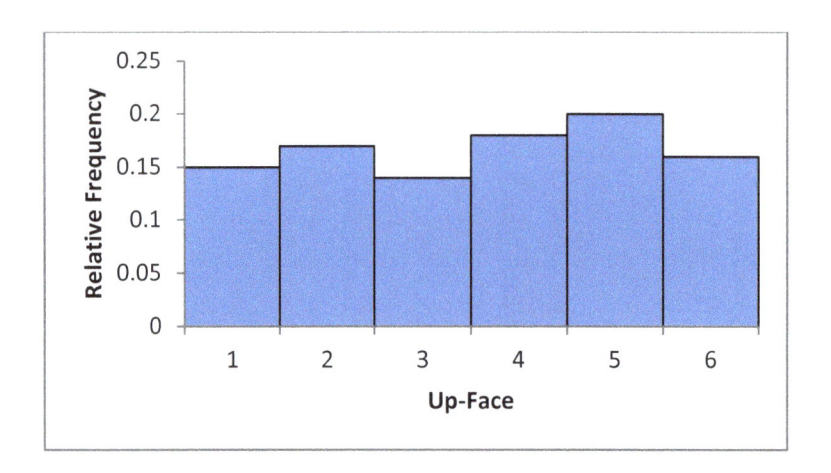

Mean: _____   Median: _____

2. The histogram of the distribution of shoe sizes of MBA seniors.

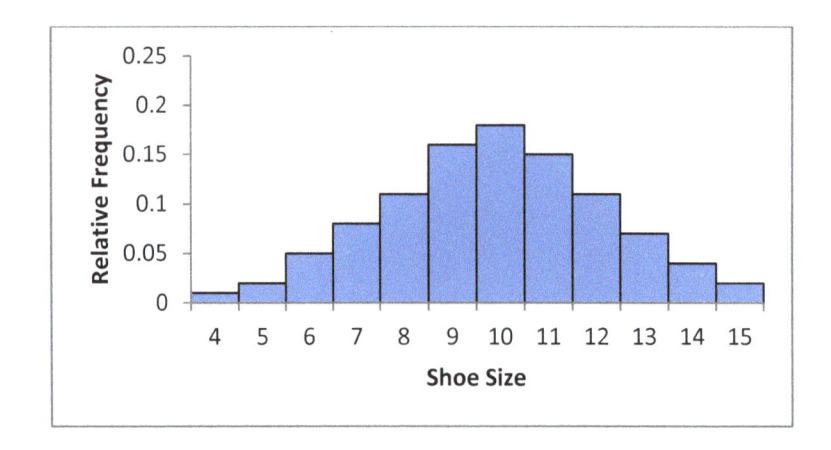

Mean: _____   Median: _____

3. A histogram of distribution of class sizes at MBA.

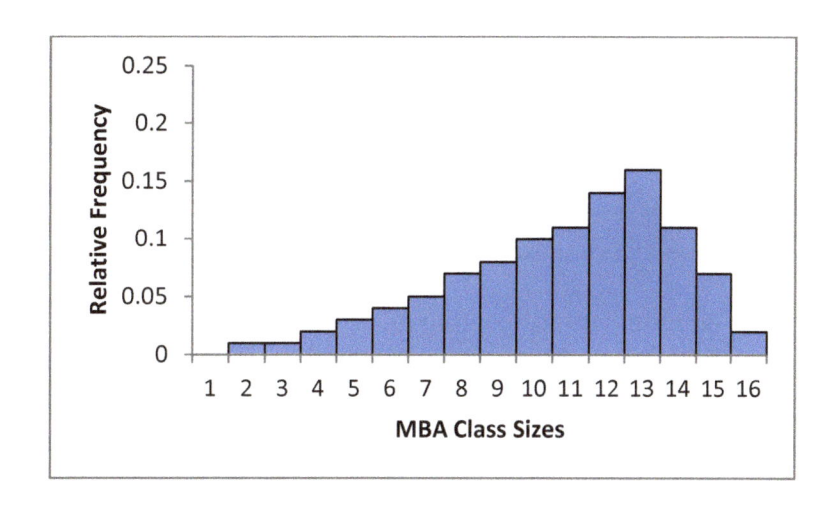

Mean: _____   Median: _____

Class Activity 5.5: Normal and Uniform Distributions

Problem 3 in Homework Assignment 5.2 showed a skewed left distribution so we could examine the effect of skewness on mean and median. Problem 1 and Problem 2 helped introduce two important types of distributions, uniform and normal respectively.

A **uniform distribution**, also known as a rectangular distribution, is a distribution where every possible outcome is equally likely. Since we were rolling a fair die in Problem 1, each of the six faces occur about 1/6 of the time in the long run and thus our density curve should look something like a horizontal line.

Since the area under any density curve is 1, if the width of a uniform density is n, the height of the density curve is 1/n. Finding the area under a uniform density curve and thus the proportion of observations in a region is simply a matter of multiplying the height of the density curve by the width of the region being considered.

A **normal distribution**, also known as a Gaussian distribution, is described by a single-peaked, symmetric and bell-shaped density curve. The normal distribution does a good job of describing many real data sets, approximating the outcome of many chance processes, and will be useful when we start working with statistical inference.

A specific normal curve is completely described by its mean and standard deviation. The mean determines the center of the distribution and the standard deviation determines the shape, specifically how spread out the distribution is. If you think back to concavity in calculus, you can see any normal distribution goes from concave up, to concave down, to concave up as we look from left to right. Fun fact: the points of inflection (where concavity changes) always occur one standard deviation to the left and right of the mean! In addition to being cool this can help when trying to draw normal distributions by hand.

An astute student might notice that since we only dealt with whole number shoe sizes, all three homework problems considered discrete variables but we approximated these discrete distributions with a continuous distribution in the form of our density curves. Since we are conceding that our density curves are merely approximations, this is acceptable, but in order to approximate well, there is a technique used to help correct for the difference in distribution type. It is called, quite appropriately, a continuity correction.

Class Activity 5.6: Distributions That Are Normal

To understand what distributions are often close to normal, let's begin by looking at chance events like coin flipping. Let's say we're looking the number of heads when a coin is tossed once. The distribution would be a discrete uniform distribution as half the time you would have 0 heads after one toss (if the coin lands T) and half the time you could have 1 head (if the coin lands H). If you toss the coin twice, however, we no longer have a uniform distribution. 25% of the time we would have no heads (if the coin lands TT), 25% of the time we would have two heads (if the coin lands HH), and 50% of the time we would have one head (if the coin lands HT *or* TH). Now let's say we toss the coin 10 times. There are 1,024 (2^{10}) equally likely outcomes if you toss a fair coin 10 times. Of those outcomes, only 1 results in 0 heads (if the coin lands TTTTTTTTTT), and thus the probability of getting 0 heads is 1/1,024. There is also only one way to get 10 heads (if the coin lands HHHHHHHHHH), and thus the probability of getting 10 heads is 1/1,024. There are 10 outcomes that result in 1 head (if the coin lands HTTTTTTTTT, THTTTTTTTT, TTHTTTTTTT...), and thus the probability of getting 1 head is 10/1,024. There are 252 ways to get 5 heads (if the coin lands HHHHHTTTTT, TTTTTHHHHH, HTHTHTHTHT, THTHTHTHTH, HHTTTHTTHH...). The most likely result is 5 heads because there are more outcomes that correspond to that number than any other and 4 heads and 6 heads are tied as the second most likely result. Even with ten tosses, the discrete distribution starts to resemble the bell-shaped normal distribution. If you toss 20 or 50 or 100 coins instead before counting the number of heads, the distribution gets even closer to normal.

    If a large group of students had no idea what the answers were to a True-False quiz and guessed on each question (half the time guessing true, half the time guessing false), the same phenomenon would occur. If it were a one question quiz, on average half the class would guess the answer right and half the class would get the answer wrong. If it were a ten question quiz, very rarely would one student get every question right or every question wrong but many students would get between 3 and 7 questions correct. One way to visualize this phenomenon is through a device called the Galton Board, which is a lot like the game Plinko on the TV show *The Price is Right*®. Discs or balls are dropped into the center of a grid where they encounter many rows of pins that ideally send the discs or balls to the right half the time and to the left half the time. While you cannot predict the end result of an individual disc or ball, in the long run balls and discs will create a distribution very close to Normal at the bottom of the board.

    More interestingly, students do not need to be guessing at random at True-False questions for this normal distribution to result. In fact, scores on tests taken by many people (like the SAT, ACT, and IQ tests and final exams of courses taught to many people like college survey classes or Latin I at MBA) tend to result in score distributions that are close to normal. Hopefully, this makes some intuitive sense. Far more students earn SAT scores (on the 2400 scale) between 1400 and 1600 than between 600 and 800.

If repeated careful measurements of the same quantity are made, the distribution of those measurements will also be close to normal. Again, this should make some intuitive sense. If this textbook is 10.12 inches long, students most students who carefully measure it with a ruler will get close to that amount (plus or minus an eighth inch, say), with a few students falling outside of that range.

The last and potentially most useful distribution that is often close to normal is characteristics of biological populations, e.g. wheat yield, turtle life expectancy, or height of an adult male human. Again this should make sense in terms of what you have observed, i.e. you have likely seen more adult males between 5'4" and 6'4" than adult males between 4'4" and 5'4" or between 6'4" and 7'4". With a large enough sample of MBA students, you really would expect the distribution of heights to look like that in Class Activity 5.4.

Class Activity 5.7: Is it a Normal Distribution? Are You Sure?

While many distributions are close to normal, a statistician needs to be careful when making assumptions about a distribution. Consider the case of Justin Wolfers, the Wharton professor who in 2006 claimed that in about 6% of all college basketball games involving a strong favorite there was a team that was willing to manipulate their performance which means that, since about 1/5 of college basketball games involve strong favorites, about 1% of all college basketball games involve gambling-related corruption.

To understand Wolfers' claim, it is necessary to understand a little bit about sports betting. Wolfers used the example of a March 2005 game between the University of Pennsylvania and Harvard. The line was set at Penn – 14.5. This means that instead of betting on simply who would win the game (called betting the moneyline), betters could participate in spread betting. This means that a bet on Penn to "cover the spread" would only be a winning bet if Penn won by 15 points or more. A bet on Harvard would win if Harvard won *or* lost by 14 or fewer points. Bookies would often pay out $10 for every $11 bet which means that if they get the same amount of money bet on either side of the spread, it is impossible for them to have a net loss.

While college athletes can certainly be influenced financially, most players would also prefer not to lose games if they can avoid it. As a result, Wolfers identifies games with heavy favorites (which he defines as when one team is favored by more than 12 points) as the ideal situation for corruption because strong players on the favored team can "shave" points without the same risk of losing they would face when trying to fail to cover a small spread (like 3 points).

Wolfers makes the assumption that the distribution of the scoring margin should be close to normal and centered around the spread. Specifically, Wolfers focused on the fact that the distribution of scores should be close to symmetric if there is no point shaving. Wolfers claims that, with no corruption, the likelihood of the favored team winning by less than the spread should be the same as the likelihood of the favored team winning by more than the spread but less than twice the spread. For example, for games where a team is favored to win by 13 points there should be the same number of wins by the favored team with a scoring margin of 1 to 12 points as there are wins by the favored team with a scoring margin of 14 to 25 points. Wolfers found that this symmetry was present when the spread was small, but when there was a strong favorite, the favorite won by less than the spread 46.2% of the time and won by more than the spread but less than twice the spread 40.7% of the time. Wolfers claimed that this asymmetry is likely the result of point shaving and that for the 16-year sample he considered, he believed that about 500 games involved gambling-related corruption.[25]

Steven Heston and Dan Bernhardt considered games where the point spread changed before the game. If people are betting on the favored team to cover the spread, bookmakers will often increase the spread in the hopes of getting that even balance of money bet on both sides of the spread that ensures them a profit. On the other side, if people are betting on the underdog, the bookmakers will decrease the spread. Heston and Bernhardt argued that, if corruption was present, you could expect large bets to be placed on the underdog by the corrupt agency before the game which would decrease the spread. Heston and Bernhardt found that, regardless of the change in the spread, the ~6% asymmetry that Wolfers noted still existed. Then, the two considered Division I basketball games in which no betting line was set. They determined lines independently an again observed about a 6% asymmetry in the results. They hypothesized that the distribution would not be normal even if corruption existed in no form and that the asymmetry was likely the result of some other factor. One of their simplest and most appeal theories was that, when a strongly favored team (say 15 point favorites) gets a significant lead (say 10 points) near the end of the game, they will likely start holding the ball to use up the shot clock and delay the number of possessions the losing team has to mount a comeback. If a less favored team (say 5 points) has a small lead (say 2 points), this strategy would be less common because even a single possession could turn the losing team into the winning team. This and other factors they identify might explain the asymmetry instead of Wolfers claim of corruption in college basketball.[26]

Homework Assignment 5.3: Is it a Uniform Distribution? Are you Sure?

For homework, we are going to collect some data and examine two distributions. Ask 20 people for the last digit in their cell phone number (if their number is (615) 555-2358 the last digit would be 8) and for the first digit of their home address (if they live at 732 Fake Street then the first digit would be 7). Please do not make up your data and please avoid asking the same people that everyone else in the class is asking (if everyone asks the same 20 people in the Senior Room our data will not be very useful). Before you collect your data, predict what the two distributions will be.

Circle One

Last Digit of Cell Number:

Approximately Normal

Approximately Uniform

Something Else

Circle One

First Digit of Address:

Approximately Normal

Approximately Uniform

Something Else

Last Digit of Cell Number

0	1	2	3	4	5	6	7	8	9

First Digit of Home Address

1	2	3	4	5	6	7	8	9

Class Activity 5.8: First Digits and Final Digits

Let's collect data from the class and see what we discover about the first digits of people's home addresses and the final digits of people's cell phone numbers.

Last Digit of Cell Number

	0	1	2	3	4	5	6	7	8	9
Total										
Proportion										

First Digit of Home Address

	1	2	3	4	5	6	7	8	9
Total									
Proportion									

Now let's make a histogram of these distributions.

Last Digit of Cell Number

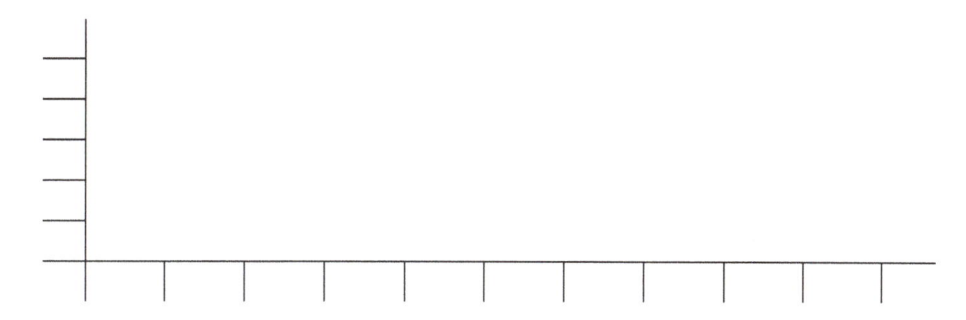

First Digit of Home Address

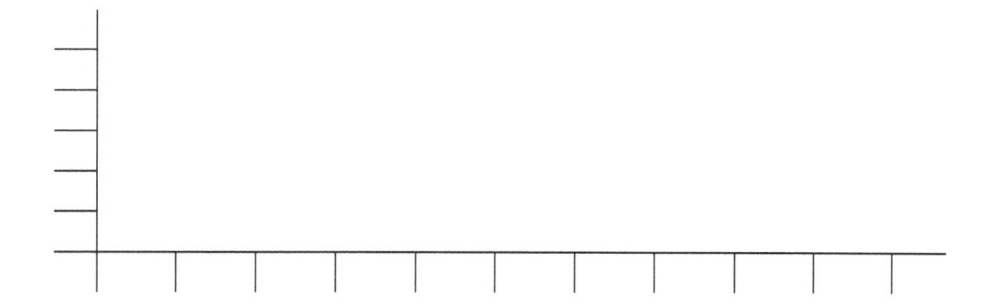

Describe the distribution of last digits of cell phone numbers:
_____

Describe the distribution of first digits of home addresses:
_____

Class Activity 5.9: The 68-95-99.7 Rule

We have already seen that the normal distribution is almost always an approximation and in some cases, might be a poor one. If that's the case, why do we use it so often? One reason is that, with normal distributions, it is very easy to determine the proportion of observations that fall between two values. One way this is evident is with the **68-95-99.7 rule**, which is also known as the empirical rule. The 68-95-99.7 rule says that in any normal distribution, approximately 68% of observations fall within one standard deviation of the mean, approximately 95% of observations fall within two standard deviations of the mean and approximately 99.7% of observations fall within three standard deviations of the mean. While this is only an approximation, it can be used on any approximately normal distribution, from IQ scores to the length of cicadas.

The height of US adult males is one of these approximately normal distributions that we can consider. The mean height of an adult US male is about 5'10" or 70 inches and the standard deviation is about 3 inches. Figure 5.1 is a density curve that describes this distribution.

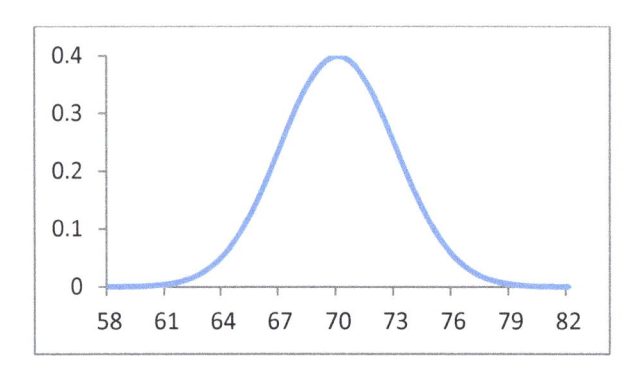

Figure 5.1 Distribution of US Adult Male Height

Using the 68-95-99.7 rule…

Between what values do the heights of the middle 95% of US adult males lie?
_____

What percent of US adult males are between 67 and 73 inches tall? ____

What percent of US adult males are between 70 and 73 inches tall? ____

If a US adult male is 73 inches tall (6'1") then he would be at what percentile in this distribution? ____

What percent of US adult males are taller than 79 inches? ____

In November of 2013, the New York Times published an article about the likelihood of making it into the NBA depending on your height, race, household income and other family traits. It claimed that there is a 1 in 1.2 million probability of making it into the NBA if you are an American male who is shorter than 6 feet tall. If you are an American male who is 7 feet or taller, there is 1 in 7 probability of making it into the NBA.[27] Part of this difference clearly comes from the fact that being taller is an advantage in basketball and that NBA teams are unlikely to play a center who is shorter than 6'9". What is the other reason why a player taller than 7 feet tall has a much better chance of making it to the NBA?

_____

If the distribution of adult American male height were approximately uniform from 4'4" to 7'4" and the population of adult American males stayed the same, how do you think that change would affect the likelihood of making it into the NBA if you were more than 7 feet tall?

_____
_____

Homework Assignment 5.4: Examining the "New SAT"

In spring of 2016, the College Board released a new version of the SAT. The previous version had three sections: *Critical Reading*, *Math* and *Writing*. The new version has two sections: *Evidence-Based Reading and Writing* and *Math*. The three sections on the old SAT and the two sections on the new SAT are all scored on a 200-800 point scale. The mean on each section is approximately 500 and the standard deviation is approximately 100. This means that the following density curve, Figure 5.2, could be used to visualize an approximation of the distribution of scores on the math section both on the old version and new version of the SAT.

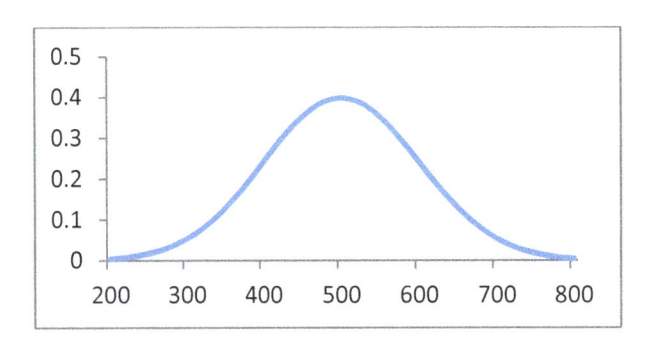

Figure 5.2 Distribution of SAT Math Section Scores

Using the 68-95-99.7 rule...
Between what values do the heights of the middle 68% of Math SAT section scores lie?
_____

What percent of Math SAT section scores are between 500 and 700? ____
What percent of Math SAT section scores are between 300 and 400? ____
At what percentile in this distribution would a 400 be? ____

The College Board posts percentile ranks for each score online. In 2014, the percentile rank for a 400 on the Math Section is exactly what you would predict using the 68-95-99.7 rule with an assumed mean of 500 and standard deviation of 100.[28] In most cases, however, this will not be the case and it is important to remember that, while it is a useful tool, the 68-95-99.7 rule only yields approximations.

The maximum possible score on a section of the SAT is an 800 which is 3 standard deviations above the mean if the mean of the distribution is 500 and the standard deviation is 100. What if the SAT did not put a cap on its high scores but instead allowed people to earn 810? 840? 950? Use the 68-95-99.7 rule to approximate the proportion of the population who would earn a score above 800 if it were possible. _____

Class Activity 5.10: The Standard Normal Distribution

While the 68-95-99.7 rule is a nice quick trick for finding some proportions, it is easy to see that its application is limited. If an adult male in the US is 72 inches tall or someone earns a 560 on the Math Section SAT, the 68-95-99.7 rule is not all that helpful. We need to figure out what to do when dealing with observations that are not exactly 0, 1, 2, or 3 standard deviations away from the mean.

The good news is that an observation's percentile in a normal distribution can be determined from its z-score. It does not matter if it is a giraffe with a height 1.3 standard deviations above the mean, an IQ score that is 1.3 standard deviations above the mean or the number of heads that resulted when I flipped my lucky fair coin 10,000 times being 1.3 standard deviations above the mean. In all three of those situations the observations would be at the same percentile level, meaning there would be the same number of observations in the distribution at or below that observed value. In these cases about 90.32% of giraffes would be shorter, about 90.32% of IQ scores would be lower and about 90.32% of the time I flipped my coin 10,000 times I could expect to have fewer heads result.

When solving problems involving normal distributions, we often use what is called the **standard normal distribution**. The standard normal distribution is the normal distribution with mean zero and standard deviation 1. It is helpful because when any normal distribution is standardized, i.e. when observations are changed into z-scores, the result will be the standard normal distribution.

At the end of this book there is a group of tables. The first is Table A which gives the area under the curve to the left of a given z-score. This table only gives areas corresponding to z-scores between -3.0 and 3.0 going up by tenths. Many standard normal probability tables will give a wider range and have more specific z-scores, but when we are looking for more exact results, we will simply use our calculators.

In this table you can see where the 90.32% came from in the last paragraph. If you earn a 120 on an IQ test and look for that score in Table A, you will be out of luck, but if you convert that score into a z-score, it rounds to 1.3 (the mean IQ score is 100 and standard deviation is 15). This z-score of 1.3 corresponds to a proportion of .9032 to the left of that z-score, or 90.32% of observations in the distribution being less. The giraffe's height and number of coin tosses that came up heads would not correspond to the same x value, but they would correspond to the same standard variable z.

In a sample, the symbol for mean is $\bar{x}$ and the symbol for standard deviation is s. In a population, the symbol for mean is μ and the symbol for standard deviations is σ.

Therefore when we discuss a normal distribution we use μ and σ. Thus, for any variable x that has a normal distribution with mean μ and standard deviation σ, the standard variable

$$z = \frac{x - \mu}{\sigma}$$

has the standard normal distribution.

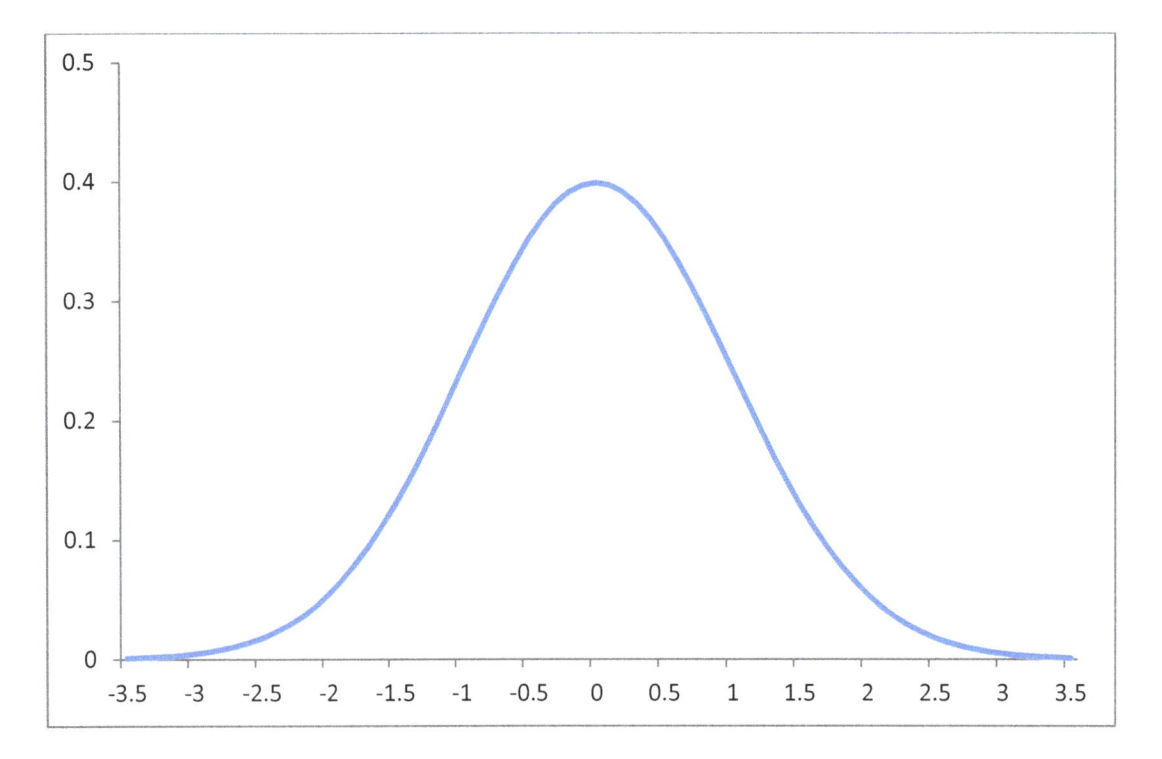

Figure 5.3 The standard normal distribution

Class Activity 5.11: Brandt Snedeker at the Driving Range

When Brandt Snedeker hits his driver, the distance the ball travels follows a normal distribution with mean 288 yards and standard deviation 10 yards.[29] This means that the distribution of Brandt's drives looks like the following.

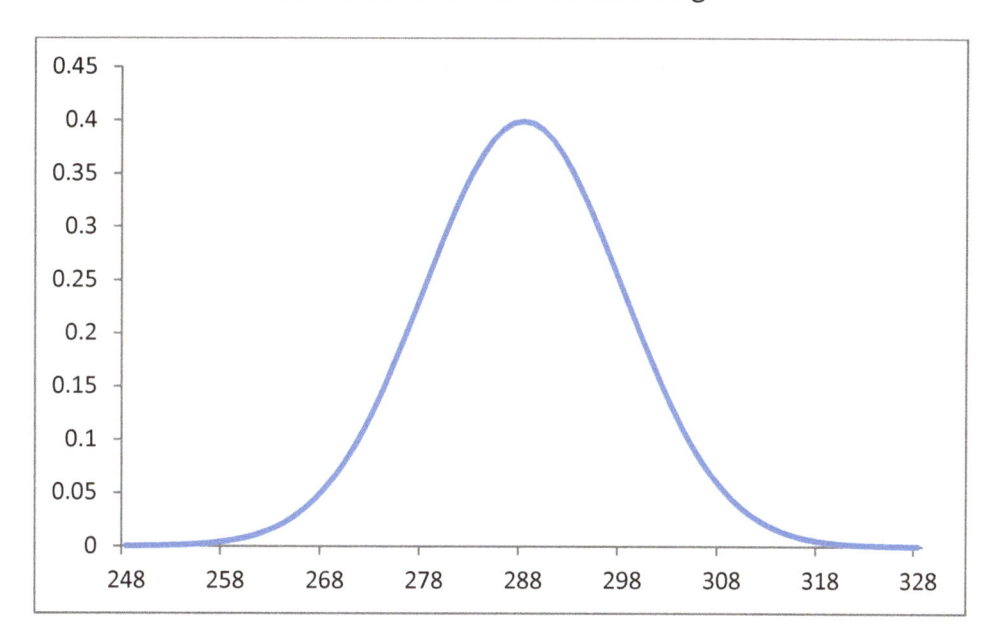

Warm-Up Question: What percent of Snedeker's drives would fall between 278 and 298 yards? _____

Using Table A, what percent of Snedeker's drives would fall between 280 yards and 300 yards? _____

Let's compare that answer to the answer our calculator gives us. To find the proportion of observations that fall in a region on a normal distribution we use normalcdf. Find the normalcdf command by pressing 2ND, VARS, 2. Then type the lower bound, the upper bound, the mean of the distribution and then the standard deviation of the distribution. When you finish your screen should look like the following:

```
normalcdf(280,30
0,288,10
```

Press ENTER to calculate the proportion of observations within that range.

What percent of drives did the calculator determine would fall between 280 and 300 yards?
_____

Use Table A to find what distance a drive would have to travel to be at the 90th percentile of Snedeker's drives.

Let's compare that answer to the answer our calculator gives us. To find the an observation that would be at a given percentile in a distribution, we use the invNorm commend. Find the invNorm command by pressing 2ND, VARS, 3. Then type the proportion that should be to the left of the observation, the mean of the distribution and then the standard deviation of the distribution. When you finish your screen should look like the following:

```
invNorm(.9,288,1
0
```

Press ENTER to calculate the drive that would be at the 90th percentile of Snedeker's drives.

What drive distance did the calculator determine? _____

Homework Assignment 5.5: Some Proportion Problems

What score does a student have to earn on the math SAT (mean 500, standard deviation 100) in order to be in the top 10%? _____

Let's say that the distribution of MBA senior heights is approximately normal with mean 68 inches and standard deviation 3 inches and the distribution of Harpeth Hall senior heights is approximately 64 inches with standard deviation 4 inches.

What percent of MBA seniors is between 59 and 77 inches tall? _____

What percent of Harpeth Hall seniors is shorter than the average MBA senior's height? _____

What percent of MBA seniors is taller than 6 feet? _____

How tall would a Harpeth Hall senior have to be to be in the 97th percentile for height in her class? _____

And now for the tough one: Let's say that, to eliminate the stress of asking someone to prom, MBA and Harpeth Hall start a new policy of randomly selecting MBA seniors and Harpeth Hall seniors to be dates for prom. If an MBA senior and a Harpeth Hall senior are selected at random, what is the probability that the girl will be taller than the boy? _____

Note: It is not possible for you to answer this question with the methods you have learned so far in the class. Try your best to figure out an answer or at least a good guess but do not spend more than 5 minutes working on this problem.

Class Activity 5.12: The Prom Date Problem

Hopefully, the first four homework problems were not too challenging and that you were able to at least put together a reasonable guess for the fifth and final homework problem. We are going to finish this chapter by talking about two methods you could use to approach a challenging problem like the prom date question.

The first approach is going to be a simulation. We will do more simulations in chapter 7, but the idea behind a simulation is that the more you run a simulation, the more confidence you can have that the overall results of a simulation reflect the underlying probabilities present in the problem. In other words, in this problem if we can simulate selecting many girls at random from the distribution described and many boys at random from the distribution described, and then see how often the girl is taller than the boy, we can be fairly confident that the actual likelihood of a girl being taller than a boy in this problem will be close to the proportion of times a girl was taller than a boy in the many trials of the simulation.

To run our simulation we are going to use the lists in our TI-83 or 84. Push STAT and then ENTER to select 1:Edit… If you have any values in your lists (L1, L2, L3 etc.), move your cursor to the title of the list and hit CLEAR to get rid of any existing data. When you've done this your screen should look like the following.

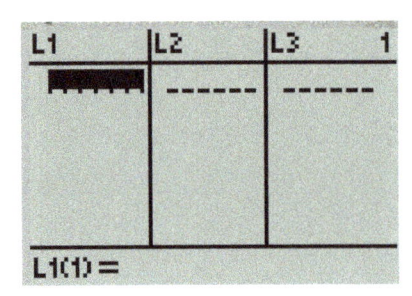

Now we want to have the calculator generate the heights of 100 Harpeth Hall seniors selected at random from the distribution described in the problem. To do this we will use the randNorm command. Use the arrow keys to select L1 and then hit the ENTER key. Then press MATH and then move over to the PRB column to select 6:randNorm(. After randNorm(is on the screen you should type 64 and then 4 to specify the mean and standard deviation of the distribution and then type 100 to say that you want to select 100 individuals at random. Your screen should look like the following.

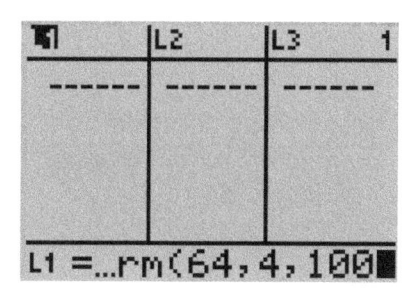

When you hit ENTER, 100 values should be entered into list 1 which represent the heights of 100 randomly selected Harpeth Hall seniors. Now we can do the same thing for list 2 to generate the heights of 100 randomly selected MBA seniors. Use the arrows to select L2, hit ENTER, and then type randNorm(68,3, 100). After you hit ENTER, you should have a list of 100 Harpeth Hall heights in L1 and 100 MBA heights in L2 and your screen should look something like the following.

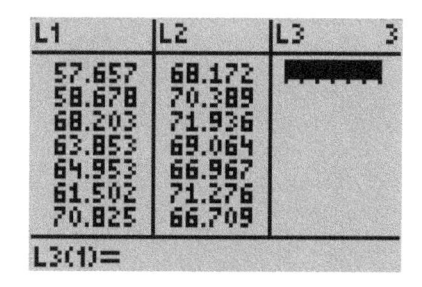

To see the difference in height for each pair, select L3, hit ENTER and type L1 – L2. Note: L1 is selected by pressing 2nd, 1 and L2 is selected by pressing 2nd, 2. If the L3 value is negative, the boy is taller. If the value is positive, the girl is taller. You could manually count how many values are positive, but there is a faster method. Select L4, hit ENTER, and type L3>0. L3 is selected by pressing 2nd, 3 and the > symbol can be selected after hitting 2nd, MATH. The equation L3>0 returns a 1 every time the statement is true, i.e. every time the girl is taller than the boy and returns a 0 every time the statement is false. Your calculator should look something like the following now.

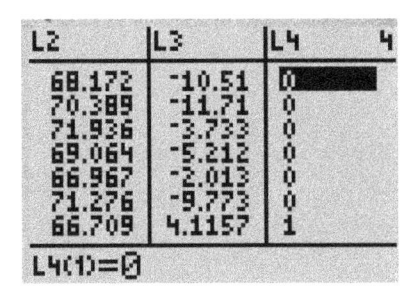

To find the number of couples with a taller girl, select 2nd, MODE to quit back to the home page. Then select 2nd, STAT, move over to MATH and select 5:sum(. When sum(is on your home page, enter L4, and when your screen looks like the following, hit ENTER.

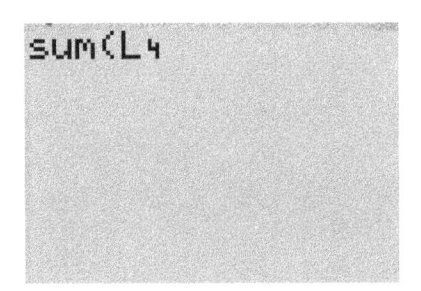

Let's see what everyone's result was from this simulation:

Student #1		Student #9	
Student #2		Student #10	
Student #3		Student #11	
Student #4		Student #12	
Student #5		Student #13	
Student #6		Student #14	
Student #7		Student #15	
Student #8		Student #16	

Based on the results of this simulation, what do you now think is the probability of a randomly selected Harpeth Hall senior being taller than a randomly selected MBA senior? _____

     Our second approach to solving this problem is going to give us a more exact answer but is going to require a little more knowledge about normal distributions. The first piece of information needed is that when the normally distributed variables are added together or subtracted, the result is a new normal distribution. If the distributions are added, the new normal distribution has a mean that is the sum of the means of the two combined distributions. If the distributions are subtracted, the new normal distribution has a mean that is the difference of the means of the two original distributions. Regardless of whether the distributions are added or subtracted, the variance (reminder, this is the standard deviation squared) of the new normal distribution is equal to the sum of the variances of the two original distributions. Using this information, find the mean and standard deviation of the distribution of girl height – boy height and find the probability that the distribution is greater than 0 using Table A and normalcdf.

According to Table A, what is the likelihood that a randomly selected Harpeth Hall senior is taller than a randomly selected MBA senior? _____

According to normalcdf, what is the likelihood that a randomly selected Harpeth Hall senior is taller than a randomly selected MBA senior? _____

ESSAYS

1) THE HOT HAND

2) ~~Chance Me~~ THE MYTH OF CHANCE MEETING

3) WINNING STATE LOTTERY 2X

4) THE MYTH OF LAW OF AVERAGES

Jaungle Juice
Chimmy Bumpis
Jumpis
Pumpis

Chapter 6. Bivariate Data Analysis

Class Activity 6.1: Our First Scatterplot: Height vs. Wingspan

So far we have only been concerned with univariate data, meaning that our data has involved a single variable like height. In this chapter, we will start working with bivariate data, which as you might guess, concerns two variables. To start, we will look at two quantitative variables, height and wingspan, measured on the same individuals, the members of the class.

Let's pair up and measure our heights and wingspans in inches and record the results in the table below.

	Height	Wingspan		Height	Wingspan
Student #1			Student #9		
Student #2			Student #10		
Student #3			Student #11		
Student #4			Student #12		
Student #5			Student #13		
Student #6			Student #14		
Student #7			Student #15		
Student #8			Student #16		

Much like when we were working with univariate data, it can be tough when working with bivariate data to interpret results without visual representations of the data or numerical values that summarize relationships between variables. The most common way of displaying bivariate data is a **scatterplot**. A scatterplot plots invididuals in a data set with each individual's horizontal coordinate representing one variable value and its vertical coordinate value representing the other.

Make a scatterplot of these data on the next page. Write a title, choose an appropriate range of values, label the hash marks, label the axes, and plot every individual in the class. In this scatterplot, there is likely a relationship between the two variables, but it is not the case that one is causally responsible for the other. That is, while taller people tend to have greater wingspans and vice versa, it would be incorrect to say that wingspan is caused by height. If, however, you think there might be some plausible causal connection between the two variables studied, for example if we collected data on hours spent studying for the math final exam vs. grade on the final exam, there is a plausible causal connection there. In those cases, the variable thought to be responsible for causing changes is called the **explanatory variable**, or independent variable. The variable that responds to changes in the explanatory variable is known as the **response variable**, or dependent variable. If there is an explanatory variable and response variable, the explanatory variable should be plotted as the horizontal value and the response variable should be plotted as the vertical value. In this case, since there is no explanatory and response variable, either option would be appropriate, but to help us compare, plot height on the x-axis and wingspan on the y-axis.

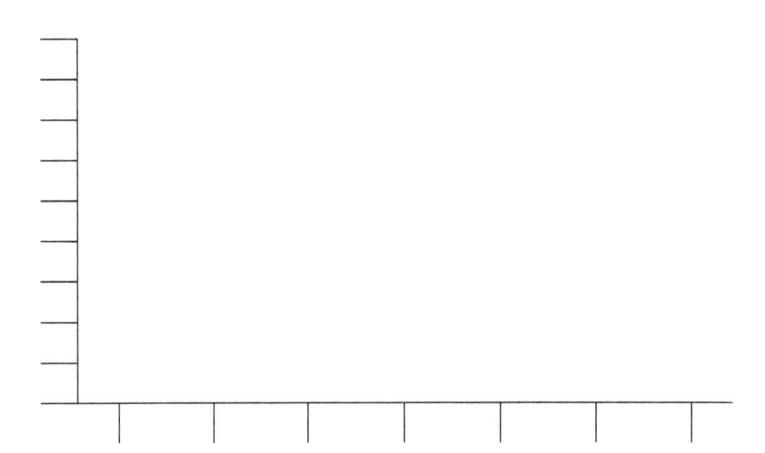

Now add Mr. Davidson to the scatterplot. He is 80" tall and has an 80" wingspan. There is a good chance that he will be outside of the horizontal and/or vertical range that you chose. While he is likely an outlier when considering height or wingspan separately, he is unlikely an **outlier** when considering the two variables together. An outlier in a bivariate data set is an individual that falls outside the overall pattern of the group. While Mr. Davidson has a greater height and wingspan than most, his height is about what you would expect for someone with his wingspan and his wingspan is about what you would expect for someone of his height.

Which student is the best candidate for being an outlier in this data set? Why?
_____
_____

When describing a scatterplot like this, we typically consider three ways of describing the overall pattern of the graph: **direction**, **form**, and **strength**. The direction of a scatterplot tells us in what direction points on the scatterplot are moving as horizontal values increase. We say that two variables have a **positive association** when above average values of one variable tend to accompany above average values of the other and below average values of one variable tend to accompany below average values of another. We say that two variables have a **negative association** if above average values of one variable tend to accompany below average values of the other and vice versa. The form of a scatterplot is what type of curve we find in the points on the scatterplot (linear, curved etc.) and whether the points occur consistently or in bunches or clusters. Lastly, the strength of a scatterplot indicates how clearly the points follow a particular pattern. The fewer deviations or outliers from the pattern we find, the stronger the relationship is in the scatterplot.

Describe the direction of our height vs. wingspan scatterplot. Specifically, do the two variables have a positive, negative or no association?

_____

_____

Describe the form of our height vs. wingspan scatterplot, e.g. linear, curved, single-clustered, double-clustered etc.

_____

_____

Describe the strength of our height vs. wingspan scatterplot, e.g. weak, moderate, strong etc.

_____

Class Activity 6.2: Distance to School and Commute Time

Let's collect some data. Share your commute to school in miles and the average number of minutes it takes you to get from leaving your house to arriving at MBA in the morning.

	Miles	Time		Miles	Time
Student #1			Student #9		
Student #2			Student #10		
Student #3			Student #11		
Student #4			Student #12		
Student #5			Student #13		
Student #6			Student #14		
Student #7			Student #15		
Student #8			Student #16		

Homework Assignment 6.1: Questions about Your Morning Drive

Is it more appropriate to think of distance and commute time as an explanatory variable and response variable or merely explore the relationship between the two?

_____

Make a scatterplot below to help us investigate the relationship between these two variables.

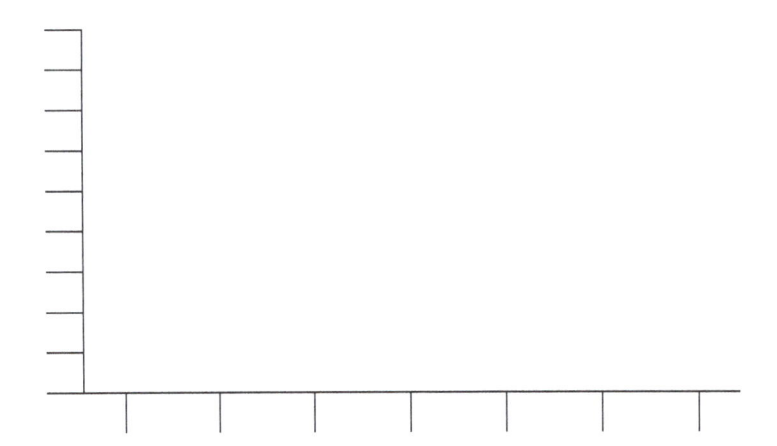

Do distance to school and commute time have a positive, negative or no association?
_____

Describe the form of the relationship.
_____

Are there any outliers? If yes, which student(s) is/are outlier(s)?
_____

Describe the strength of the relationship. Is it weak, moderate or strong?
_____

Class Activity 6.3: Finding Correlation

So far we have measured the strength of the relationship between two variables by looking at a scatterplot and judging whether we thought it was weak, moderate or strong. These are subjective definitions and we are not always good at judging strength visually. Luckily, there is a number that describes the strength of the linear relationship between two variables which we call **correlation**. To find correlation (r), we use the following formula:

$$r = \frac{1}{n-1}\sum\left[\left(\frac{x-\bar{x}}{S_x}\right)\left(\frac{y-\bar{y}}{S_y}\right)\right]$$

The following is a table that shows four MBA students and their maximum bench press and maximum squat.

Student	Bench Max (x)	Squat Max (y)
Robert	200	250
Thomas	220	260
Stephen	240	280
Philip	260	290

Let's find the correlation (r) between Bench Max and Squat Max.

First, what does n =? __4__

While you could find it yourself, I will tell you that $\bar{x} = 230$, $\bar{y} = 270$, $S_x \approx 25.8$ and $S_y \approx 18.3$.

Given those means and standard deviations, find $\left(\frac{x-\bar{x}}{S_x}\right)\left(\frac{y-\bar{y}}{S_y}\right)$ for…

Robert: 1.271

Thomas: 0.212

Stephen: 0.212

Philip: 1.271

[handwritten notes: correlation coefficient; r = measures direction & strength of a straight line relationship between two quantitative variables; r^2 = coefficient of determination]

Now let's add up those four results to find $\sum\left[\left(\frac{x-\bar{x}}{S_x}\right)\left(\frac{y-\bar{y}}{S_y}\right)\right]$: __2.966__

Finally, multiply that sum by $\frac{1}{n-1}$ to get correlation (r): __0.989__

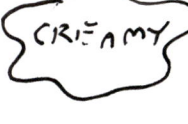

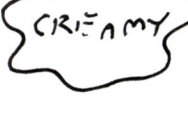

Correlation is always a number between -1 and 1 inclusive.
Were the products of the x and y z-scores positive or negative for our four individuals?
Why? __Positive – there's strong correlation__

Given your last answer, if two variables have a positive correlation, what does that indicate about the two variables? _There'd be a positive line of best fit_

What about if the two variables have a negative correlation? _There'd be a negative slope_

What if the correlation between two variables is close to zero? _There's very little correlation_

What would correlation equal if all x values or all y values were the same? _1_

Class Activity 6.4: Finding Correlation on Your Calculator

It is easy to see from our previous activity that calculating correlation by hand, especially if there are more than four individuals, can be a tedious process. As a result, we normally rely on calculators or computers to do the computation for us. Let's see how to find correlation on your TI calculator.

First, let's input our Bench Max values into L1 and our Squat Max values into L2. As a reminder, you can edit your lists (L1, L2, etc.) by hitting STAT, and then 1:Edit… After you've entered the values, your screen should look like the following.

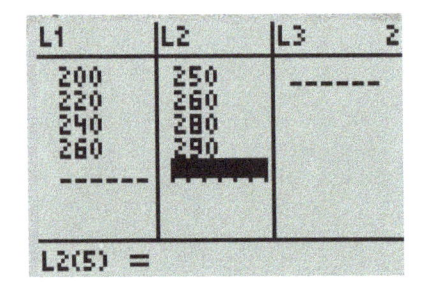

Press 2nd and QUIT to return to the home screen. Then push STAT, → to get to the CALC column and then scroll down for 8:LinReg (a + bx). Select 8:LinReg and then push ENTER to calculate. There is a good chance your screen will now look like the following.

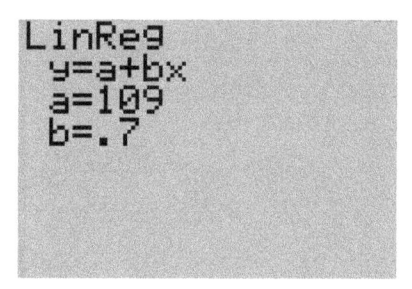

```
LinReg
 y=a+bx
 a=109
 b=.7
```

Unless your calculator has been previously used to find correlation, this is all the information the calculator will provide. To have the calculator find correlation, select CATALOG by pressing 2nd, 0. Once you are in the catalog, scroll down to DiagnosticOn, and hit ENTER twice. Now if you calculate linear regression, the calculator will also tell you correlation. If you do it correctly, your screen will look something like the following.

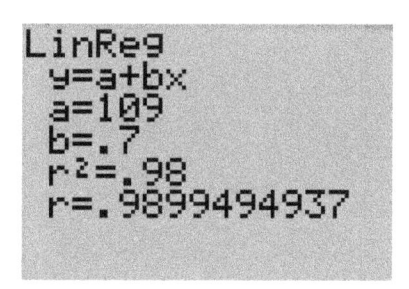

```
LinReg
 y=a+bx
 a=109
 b=.7
 r²=.98
 r=.9899494937
```

You can see that this correlation value is very close to the correlation value that we found manually. The difference comes from our rounding. You will also notice that there is another value, r^2, which the calculator provides. We will discuss r^2 later in this chapter.

Homework Assignment 6.2: Exploring Correlation (r)

The following is a table that shows five MBA students and their shoe size and height in feet.

Student	Shoe Size	Height (ft)
Anthony	6	5.25
Joseph	8	5.5
Timothy	12	6.25
Connor	11	6
Derek	14	6.5

Before doing any calculation, do you think the correlation between shoe size and height for this data set will be positive or negative? Why? _____
_____

Use your calculator to find the correlation between shoe size and height. r = _____

Now switch the x values and y values, e.g. if you put shoe size in L1 and height in L2, put height in L1 and shoe size in L2. Find correlation.

r = _____ Why do you think you got this result? _____
_____

These height measurements were taken with shoes off. If the boys kept their shoes on, their heights would be .1 feet greater each and the table would look like the following.

Student	Shoe Size	Height with shoes on (ft)
Anthony	6	5.35
Joseph	8	5.6
Timothy	12	6.35
Connor	11	6.1
Derek	14	6.6

Find the correlation of shoe size vs. height with shoes on. r = _____

Does this result surprise you? Why? _____
_____

Lastly, instead of using feet we could have measured height in inches. If so, the original table would have looked like the following.

Student	Shoe Size	Height (in)
Anthony	6	63
Joseph	8	66
Timothy	12	75
Connor	11	72
Derek	14	78

Find the correlation of shoe size vs. height in inches. r = _____

Does this result surprise you? Why? _____
_____

Class Activity 6.5: Understanding Least-Squares Regression

So far we have discussed correlation, which is a value that measures both the strength and direction of a linear relationship. Now we will turn our attention to **regression**, which is the process of drawing a line in an attempt to describe the relationship between variables.

Consider the scatterplot below. Draw a line that you think does a good job capturing the overall trend of the scatterplot.

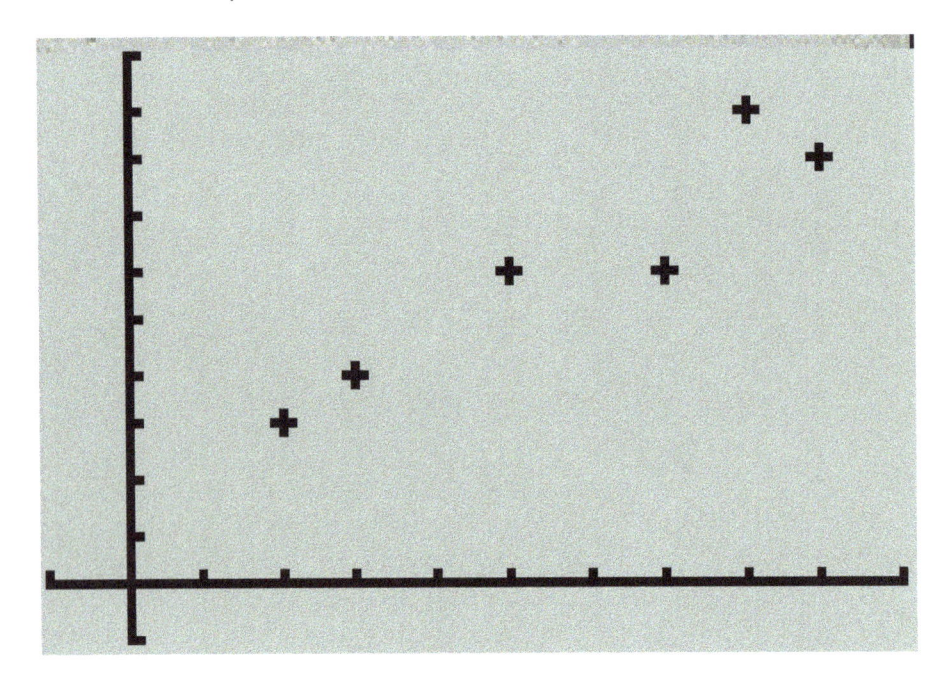

There are a lot of approaches to drawing this "line of best fit." You might notice that you can draw a line that goes through four points perfectly but misses two points considerably or you might draw a line that has three points above the line and three points below. To understand the most common regression method, **least-squares regression**, follow these steps.

1) Draw a vertical line segment from every point to the line that you drew on the graph above.

2) Turn every line segment into the side of a square. Draw the six squares. If your line goes perfectly through a point, just make a point to represent a square with no area.

When the sum of the areas of those squares in minimized, you have found the least-squares regression line. You can calculate the equation of this regression line by hand but it is, like finding correlation, very tedious. As a result, it is more common to rely on a calculator or computer.

Class Activity 6.6: Doing Least-Squares Regression on a Calculator

We actually found the least-squares regression line when finding correlation in Class Activity 6.4. We will make only one significant change when finding the least-squares regression line in this activity. Consider the following table which looks at a house in Franklin, TN for 10 months and compares the average temperature in Fahrenheit to the gas bill rounded to the nearest dollar that month.

Month	Average Temperature (F)	Gas Bill (Dollars)
January	38	64
February	42	55
March	50	47
April	61	38
May	68	30
June	76	22
July	80	17
August	78	18
September	73	30
October	60	37

First, let's input our Average Temperature (the explanatory variable) values into L1 and our Gas Bill (the response variable) values into L2. Recall that you can edit your lists (L1, L2, etc.) by hitting STAT, and then 1:Edit... After you've entered the values, your screen should look something like the following.

Press 2nd and QUIT to return to the home screen. Then push STAT, → to get to the CALC column and then scroll down for 8:LinReg (a + bx). Select 8:LinReg, then finish the command with L1, L2, Y1. This finds the least-squares regression line with L1 as the x-values and L2 as the y-values, then inputs the resulting equation into Y1 for you. The comma is above 7 and to get Y1 press VARS, hit the right arrow to get to Y-VARS, and then hit ENTER twice to select 1:Function and then Y1. Once your screen looks like the following...

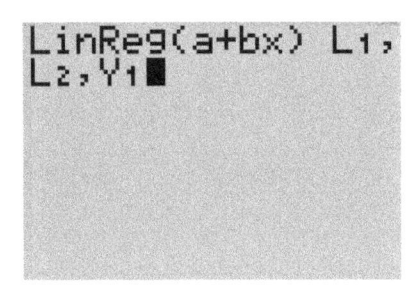

push ENTER to calculate. Assuming you have not changed your settings on linear regression, the calculator should give you correlation and r^2 in addition to the y-intercept and slope of the regression line like this:

Note that we use the slope-intercept y = mx + b in algebra but y = a + bx is more common in statistics. If you press Y=, you will see that this regression equation has been stored for you as Y1. Let's take a look at a scatterplot of these values along with the regression line. Press 2nd, STATPLOT, select the first plot and set up the plot like the following:

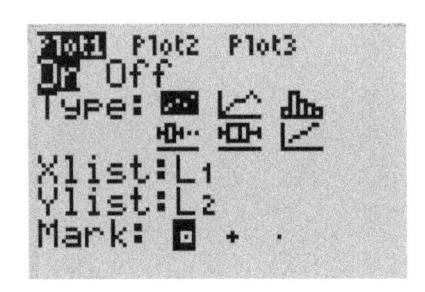

Press ZOOM, and then 9 for 9:ZoomStat to capture the scatterplot and regression line. Your screen should look like the following:

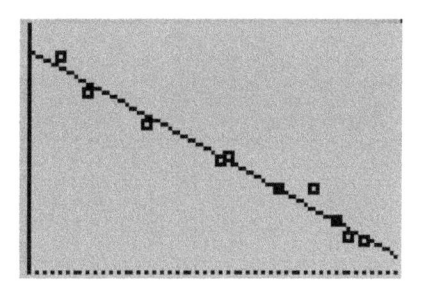

What is the slope of the regression line? What does its value mean in terms of temperature and the gas bill at this Franklin home? _____
_____

What is the y-intercept of the regression line? What does its value mean in terms of temperature and the gas bill at this Franklin home? _____
_____

One use of regression is prediction. If the average temperature in November in Franklin over the past 100 years has been 48 degrees Fahrenheit, what would you predict the gas bill will be in November for this home? _____
_____

How confident should you be in this prediction? Why? _____
_____

It is risky to make predictions outside the range of available data. For example, in this case it would normally be considered bad practice to predict a gas bill for a month where the temperature is greater than 80 or less than 38 degrees. This kind of prediction is known as extrapolation. To help see why this can be risky, predict the gas bill if there is a heat wave and the average temperature one month is 100 degrees Fahrenheit. ____

Homework Assignment 6.3: Exploring Regression and Prediction

Let's take a look at average points scored per game and average points allowed per game by the 11 Tennessee Division II-AA private school football teams during the 2014 season.[30]

Team	Average Points Scored Per Game	Average Points Allowed Per Game
McCallie	41.7	28.2
MBA	36.9	23.5
Brentwood Academy	36.9	25.7
Ensworth	36.0	20.0
MUS	29.5	20.2
Father Ryan	25.8	17.7
JP2	24.1	40.8
Saint Benedict	23.6	23.0
Baylor	23.3	15.6
Briarcrest	21.4	12.6
Christian Brothers	13.7	8.1

Put Average Points Scored Per Game into L1 and Average Points Allowed Per Game into L2 and then calculate the least-squares regression line and the correlation between the two variables and store the equation as Y1.

Equation: _____   Correlation (r): _____

Press STAT, → for CALC, and then 2: 2-Var Stats. Hit ENTER to get some summary statistics on L1 and L2. What are $\bar{x}$ and $\bar{y}$ (the mean x value and mean y value)?

$\bar{x}$: _____ $\bar{y}$: _____

Make a scatterplot of L1 and L2 and then press Zoom, 9 for 9: ZoomStat to see the scatterplot and least-squares regression line. Press 2nd, Calc, 1:Value and find the value of y on the regression line when x = $\bar{x}$. What does y =? _____

Given this result, what do you think might be a characteristic of all least-squares regression lines? _____

Of the 11 teams in the scatterplot, which one is the outlier? _____
Delete that team from the data and then recalculate correlation (r) for the 10 remaining teams.

r = _____

What does this change in correlation tell you about r and its response to outliers?

_____

Class Activity 6.7: Understanding the Coefficient of Determination (r^2)

When we found correlation for the first time on our calculators, the calculator provided another value, r^2, in addition to correlation. This value, known as the **coefficient of determination**, is the proportion of variation that can be explained by the least-squares regression line of y on x. While there are a number of formulas that can be used to calculate r^2, the simplest is to just square the correlation value r. This means that if r is 1 or -1, then r^2 is 1 and 100% of the variation in y can be explained by its linear relationship with x. If r = .8, r^2 = .64 and if r = -.7, r^2 = .49. r^2 is always a value between 0 and 1 inclusive.

To get a better understanding of what r^2 really indicates, let's consider a scenario where we are rolling a die twice. Either using a real die or using randInt(1,6) to simulate rolling a die, record the result of the first roll in column one and the sum of the first two rolls in column two. Repeat this process 10 times:

Result of First Roll	Sum of First Two Rolls

What do you predict will be the coefficient of determination (r^2) for these data? Why?
_____
_____

On your calculator, input Result of First Roll values into L1 and Sum of First Two Rolls into L2 and calculate r^2. r^2 = _____

This was a small sample of die rolls so it is very possible that the r^2 value we find does not give a good indication of what proportion of the sum of two die rolls can really be explained by the result of the first roll. To find an r^2 value that gives a better indication of how much the sum can be explained by the first roll result, we could increase the number of trials, or we could calculate r^2 with points representing all 36 equally likely results when you roll a die two times. For the sake of ease, let's do the latter. This time put the following values into L1 and L2.

Result of First Roll	Sum of First Two Rolls	Result of First Roll (continued)	Sum of First Two Rolls (continued)
1	2	4	5
1	3	4	6
1	4	4	7
1	5	4	8
1	6	4	9
1	7	4	10
2	3	5	6
2	4	5	7
2	5	5	8
2	6	5	9
2	7	5	10
2	8	5	11
3	4	6	7
3	5	6	8
3	6	6	9
3	7	6	10
3	8	6	11
3	9	6	12

What does r^2 = _____

Class Activity 6.8: Residuals and Residual Plots

One way of determining if our least-squares regression line is capturing the relationship between two variables well is by finding the coefficient of determination (r^2). It is also useful to look at **residual** values after fitting the regression line. A residual is the vertical distance between a point and the regression line, that is, residual = observed value of y – predicted value of y. Small residual values (statisticians often find the standard deviation of the residual values to measure their spread) tend to indicate that the regression line is capturing the relationship between the two variables well. Keep in mind that, unlike r^2, residual values' largeness or smallness is context dependent. If all the residual values on a 100 point scatterplot are less than 10 pounds, that could be mean that the line has great predictive strength if we are predicting elephant weight given age, but far less if we were predicting squirrel weight given age.

Another common use of residuals is a **residual plot**. A residual plot takes individuals' explanatory variable values and plots them against their corresponding residual values. If the residual plot shows a curved pattern, that indicates that the relationship between the explanatory and response variable is not linear. If there is no obvious pattern in the residual plot, linear regression is appropriate for those variables.

Let's see how to make a residual plot and find residual values on a calculator. A new hot chicken restaurant opens in Sylvan Park. An MBA student is wondering how pricey it is, so he collects some data from people leaving the restaurant one evening. He asks if they would be willing to say 1, how many people were in their party and 2, what the bill was. The following table is a collection of his results.

Number of People	Bill ($)
2	24.90
5	55.15
4	46.11
3	18.67
6	80.12
1	5.09

Put the Number of People values into L1, the Bill($) values into L2, make a scatterplot of the data, find the least squares regression line and store it in Y1. Hit Zoom, 9 for 9: ZoomStat. Your graph should look like the following.

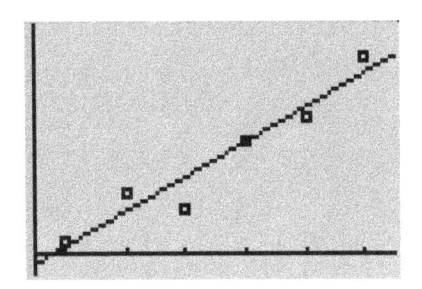

Now let's make a residual plot. Press 2nd, STAT PLOT, ENTER to select Plot 1. We're going to keep the XList as L1 but change YList to RESID. Press the down arrow until you get to YList and then hit 2nd, LIST and then scroll down for RESID. Your screen should look like the following.

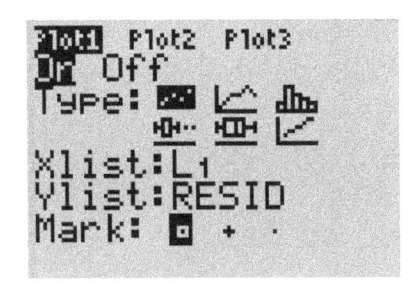

Press Y= and delete the Y1 regression line. Then press ZOOM, 9 for 9: ZoomStat. You should now see the residual plot. If you hit the TRACE button you can jump from point to point using the left and right arrow keys. Your screen should look something like the following.

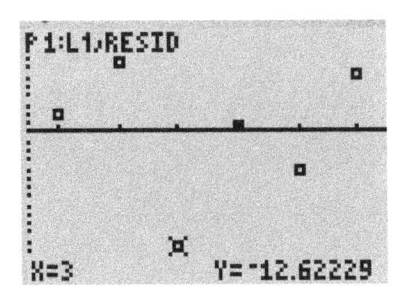

Using the TRACE option on your calculator, jump from value to value and record the residuals below. The residual for the party of 3 has already been done for you. Use as many digits as the calculator provides.

Number of People	Bill ($)	Residual
2	24.90	
5	55.15	
4	46.11	
3	18.67	-12.62229
6	80.12	
1	5.09	

What is the sum of those 6 residual values? _____

If those residual values were not approximations, what do you think the sum of the residuals would be? _____

Homework Assignment 6.4: The Coefficient of Determination vs. Residual Plots

Consider the following table with data on membership in MBA's Science Olympiad program over the years.

Year	Members
1	54
2	58
3	63
4	67
5	72
6	77
7	83
8	90
9	96
10	103

Put years into L1 and years in L2. Find the following:

Regression Equation: _____

r: _____

r^2: _____

Now make a residual plot. What do you notice about the residual values? Is this surprising? Why? _____
_____
_____
_____

Class Activity 6.9: Correlation vs. Causation

Much of chapter 6 has been spent discussing correlation. It is important to note that, while correlation is a useful tool in determining the predictive power of one variable on another, it does not necessarily imply causation. To see why, let's consider a couple situations.

For our first situation, let's say a student is looking at some data on his favorite dessert spot in Nashville, Las Paletas, and is shocked when he realizes that there is a strong positive correlation between Las Paletas's monthly sales and drowning deaths in Tennessee. Of course it is not because Las Paletas popsicles are causing people to drown, both variables are responding to a **lurking variable**, temperature. A lurking variable is a variable that is not considered in an initial study but has an important effect on the variables that are being studied. When the temperature is warm, people buy more popsicles and go swimming more often and thus there is an increased number of drowning deaths. When two variables respond to some lurking variable it is known as a **common response**.

For our second situation, let's say that Mr. Davidson decides he is going to get healthy. He cuts out red meat, processed sugar, gluten and caffeine and commits himself to exercising every day. He monitors his weight, blood pressure and cholesterol and is very happy to see that he makes steady improvements as the weeks pass. What's the problem? Mr. Davidson finds that he is very happy with his progress but does not know which of the many lifestyle changes he made were causally responsible for the change in his health. These variables are said to be **confounded**. Variables are confounded when their effects on a response variable cannot be distinguished from each other. The solution for Mr. Davidson would be to do an experiment. Instead of making five lifestyle changes at once, make one at a time and monitor the changes to his health over a period of months. *Well-designed experiments are the best means of determining causation.* This is not an issue for Mr. Davidson, but there are situations where it is not possible for ethical or practical reasons to conduct experiments. In these situations, statisticians are forced to use the data available to try to determine causal connections.

Homework Assignment 6.5: Exploring Causation

The Greater Nashville Travel Agency puts up billboards around town that say "People who take more vacations live longer." A curious statistics student contacts the company to find out what the basis is for the claim. The company says that it has been found that there is a strong positive correlation between the number of vacations a person takes a year and life expectancy. Is there a potential lurking variable that could be responsible for both the number of vacations people take and life expectancy? If so, what is it?

_____
_____
_____

If this correlation is the result of a common response, could a person use number of vacations a person takes per year to predict life expectancy?

_____
_____
_____

A teacher in his first year at MBA writes something positive on tests when a student earns a 95 or above and something negative on tests when a student earns a 75 or below. Near the end of the year, the teacher is perusing his gradebook on Scholar and notices a trend. When he writes a positive note on a test, students almost always do worse on the next test. When he writes a negative note on a test, students almost always do better on the next test. The teacher decides that students have better responses to negative feedback and resolves to be nothing but negative from this point forward. A veteran statistics teacher explains that the first year teacher's causal conclusion may be incorrect.

Regression (or reversion) **to the mean** is the idea that, after an outlier result, it is more likely that the next result will be closer to the mean than the outlier value. To understand this more fully, imagine a B student in a class. He is likely to have some test days where everything goes his way: he got a rare good night's rest, he did not have a lot of homework the night before so he had time to study, he did not miss any lectures the previous two weeks, and as a result he gets a rare 98 on the test. The student is also likely to have some days where nothing goes his way: he got 3 hours of sleep the night before, had a heavy load of homework, missed three of the lectures during that chapter because of sports and extra-curricular activities and as a result he earned a 72 on the test. After the high grade, it is likely that things won't go as well for the student on the next test and his grade will go down. After the low grade, it is likely that things won't go as poorly for the student on the next test and his grade will go down. With the first year teacher's behavior, positive comments were confounded by high test results and negative comments were founded by low test results. With the data the first year teacher has, it would be tough to conclude that it was the comments and not regression to the mean that was responsible for the changes in test grades.

You may be familiar with the "Sports Illustrated Cover Jinx" or the "Madden Curse." The idea is that players or teams who are on the cover of Sports Illustrated or Madden tend to experience bad luck soon after. What is a possible non-curse explanation for the disappointing results that athletes and teams often have after appearing on a Sports Illustrated or Madden cover? _____
_____
_____

What is an example of a situation where statisticians have had to rely on observational data because an experiment would not be possible for ethical or practical reasons?
_____
_____
_____

Class Activity 6.10: Bivariate Data Project

To conclude our chapter on bivariate data analysis, we are going to do a project instead of taking a test. First, you are going to start by choosing a population that interests you. I do not care if it is NFL players, Canadian provinces, MBA one-acts, ballpoint pens or British monarchs as long as it is a population that interests you and is a population from which you can gather or collect quantitative data.

Second, you are going to choose two quantitative variables that you think may be related in the population whose values for individuals in the population are available. You might think that there is a positive correlation between batting average and salary for professional baseball players, that there is a negative correlation between length in minutes and tickets sold for MBA one-acts or that there is no correlation between your grade in Statistics and your ACT score.

You should attempt to collect data on at least 10 individuals in such a way that avoids bias, e.g. random sampling. After you have collected your data, you should make a scatterplot of your data on your calculator, find the regression equation, find correlation and the coefficient of determination and then answer the following questions.

1) How did you try to eliminate bias when collecting data? Do you think your data is representative of the population? Why? _____
_____
_____
_____

2) Recreate the scatterplot you made on your calculator. Label the explanatory variable on the horizontal axis, the response variable on the vertical axis, and choose an appropriate range of x and y values to capture all of the data before plotting the points.

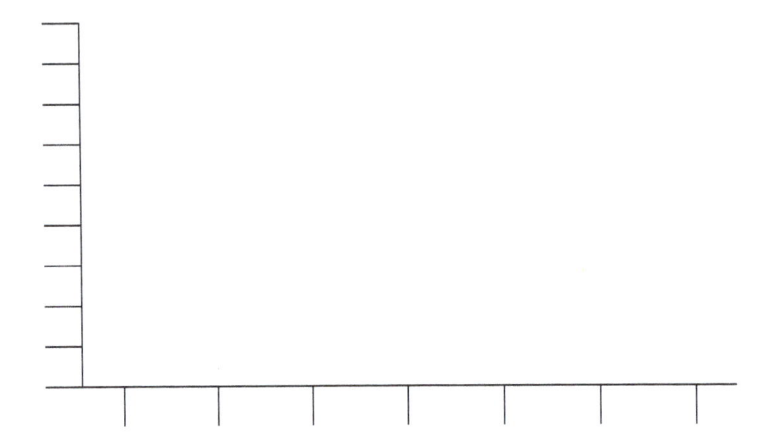

3) Sketch the least-squares regression line you found on your calculator on the scatterplot. What is the formula for the least-squares regression line?
_____

4) What do r and r^2 equal? r = _____ r^2 = _____ What do those values indicate about the relationship between these two variables? _____
_____

5) What is the slope of the regression line? What does it mean in the context of this study?
_____
_____
_____

6) Make a residual plot on your calculator and recreate the plot below.

Based on the residual plot, do you think that the relationship between the two variables is linear? Why? _____
_____
_____

7) If there is a strong correlation between the two variables, do you think that the connection is causal? Why? _____
_____
_____

Chance Behavior is regular in the long run.
Personal Probability is from 0 to 1 inclusive
Two disjoint events with probabilities > 0 are not independent
Addition Rule → $P(A \cup B) = P(A) + P(B) - P(A \cap B)$

Expected Value

- Expected value on a single die?
    - Expected Value = number × probability

$$= \sum_{n=1}^{6} np = (1)\left(\frac{1}{6}\right) + (2)\left(\frac{1}{6}\right) + (3)\left(\frac{1}{6}\right) + 4\left(\frac{1}{6}\right) + 5\left(\frac{1}{6}\right) + 6\left(\frac{1}{6}\right)$$

$$= \frac{1}{6}(1+2+3+4+5+6) = \frac{1}{6}(21) = \frac{7}{2} = \boxed{3.5}$$

- Expected value is not just 0 through 1

* SCENARIO

- Game with 4,000 scratch-offs
- 432/4,000 are redeemable

Number of tickets	4	8	20	400
Value	100	50	20	2

- Find EV of a ticket that sells for a dollar?

    - $EV = 100\left(\frac{4}{4000}\right) + 50\left(\frac{8}{4000}\right) + 20\left(\frac{20}{4000}\right) + 2\left(\frac{400}{4000}\right) + 0\left(\frac{3568}{4000}\right)$

    $= \frac{2000}{4000} = 50¢$

    - After subtracting $1.00 cost of ticket, we can expect to lose $0.50 on each ticket we buy.

Know GOP

Part C: Probability

Chapter 7. Probability Rules

Class Activity 7.1: Randomness and Probability

When many people hear the word **random** they think unpredictable. Strictly speaking, that is only half true in statistics. A phenomenon is random if an individual outcome is uncertain *but* the results of repeated trials become more predictable as the number of trials increases. This important and useful fact is guaranteed by the **Law of Large Numbers**, which says that the proportion of times a particular outcome from a chance process occurs will approach a single value if that chance process is repeated more and more times. The value approached is called the **probability**. Probability is always a number between 0 and 1 inclusive. If an event always occurs, its probability is 1. If an event never occurs, its probability is 0. If it occurs 1 out of every 4 times, its probability is .25.

Most people would be able to give a **personal probability**, or subjective perception of an event's likelihood, in some situations and many are aware of the Law of Large Numbers and of the idea of probability, but the relationship between this law and probability is frequently misunderstood. It is true that there are two sides to a coin and the probability of landing heads when a fair coin is tossed is ½, or .5. In addition, it is true that there are six sides on a die and that the probability of rolling a three on a die is 1/6, or ~.1666. However, finding probability is not always as simple as dividing by the total number of outcomes.

What is an example of a random process that has two possible outcomes but the probability of the two outcomes is *not* .5? _____
_____
_____

What is special about flipping a coin and rolling a die that makes the probability of landing heads or rolling a three ½ and 1/6 respectively?
_____
_____

One misinterpretation of the **Law of Large Numbers** is the **gambler's fallacy**. The gambler's fallacy is the mistaken belief that if a particular outcome has happened at a high frequency, it will happen less frequently in the future. For example, after someone witnesses a coin being tossed and landing heads ten times in a row, the observer believes that we are due for a tail. Or, a couple that has had four boys believes they are due for a girl. Interestingly, the opposite also happens. A person may believe that because a roulette wheel has come up red 5 times in a row, it is more likely to continue to be red. In all three cases, the observer is wrong because the trials are **independent**. Two events are independent if the outcome of one does not affect the probability of the other. If you aren't due for heads after tails, or a non-3 after a 3, how does the Law of Large Numbers work? Consider the following. You have a fair Tennessee quarter and you toss it 10 times and every time it comes up heads.

What proportion of your first ten tosses is heads? _____

What is the likelihood that your next toss will be heads? _____

If you toss the coin 9,990 more times, and half of those results are heads, how many total heads came up in your 10,000 tossses? _____

What proportion of your first 10,000 tosses were heads? _____

Improbable events are, by definition, likely to happen very few times in many trials. However, if there are many opportunities for something improbable to happen, it should not be surprising when it does. To understand this better, consider a classroom at MBA of 15 students. Assuming that there are no twins or triplets in the class and that certain birthdays aren't more likely than others, how likely do you think it is that there are two or more students with the same birthday? Would it surprise you to find out that the probability of a repeat birthday is about 1/4, or more accurately ~.2529? In fact, if a classroom had 23 or more students, it would be more likely that there was a repeat birthday than there was not. What if a class of 15 students listed their birthdays and the birthdays of their mothers and fathers. With those 45 birthdays listed, there would be about a .941 probability of a repeat. Calculating this probability is not that challenging, but what is more important is the idea behind this surprisingly high value. If the question was instead, what is the probability that a student or a parent of a student in Mr. Davidson's class has the same birthday as Mr. Davidson, the probability would be much lower, approximately .116. But, in the original question, there are so many ways for the ostensibly improbable repeat birthday to occur. The mother of the tallest kid in class could have the same birthday as the father of the shortest kid in class, the father of the fastest kid in class could have the same birthday as the father of the slowest kid in class, etc. Since there are so many ways for a repeat birthday to occur, it is not surprising when it happens.

If Mr. Davidson runs into a classmate from his high school in Vermont in a Brentwood Kroger one weekend, it might seem like an improbable event, and it is. So what happened? Fate? Kismet? Destiny? It could be, but it seems like there is a more reasonable

explanation. Mr. Davidson does stuff every day. He gets groceries. He fills up h
gets his hair cut. He gets his oil changed. He goes out to dinner. He goes shop
to sporting events. Mr. Davidson also has classmates from elementary and m
well as college. What is the probability that on a particular day Mr. Davidson
a classmate from high school at the Brentwood Kroger? Very small. What is t
that at some point this year Mr. Davidson will run into a classmate from middle school, high
school, or college at the grocery store, gas station, barber, auto shop, restaurant, mall, or
while watching sports? Much greater.

When Apple first introduced the shuffle feature on the iPod, people complained that the shuffle was far from random because they were hearing the same songs too frequently. In reality, it would be shocking if some songs played at random *didn't* play multiple times before you heard every song on your iPod playlist. In response to the complaints, Apple adjusted the feature to be less random to avoid so many repeats.[32]

Homework Assignment 7.1: Exploring The Law of Large Numbers

1. If Johnny Student rolls a fair die four times and it lands 4 every time, what proportion of the first four rolls are 4s? __100%__

2. What is the probability that the next roll comes up 4? __1/6__

3. If 1/6 of the next 96 rolls are 4s, then what proportion of the *first 100* rolls are 4s? __20%__

4. A fair coin is flipped twice. What is the probability of getting:

5. HH? __1/4__ HT? __1/4__ TH? __1/4__ TT? __1/4__

6. All heads? __1/4__ All tails? __1/4__ One head and one tail? __2/4__

7. A fair coin is flipped four times. What is the probability of getting:

8. All heads? __1/16__ All tails? __1/16__ Two heads and two tails? __3/8__

9. After how many tosses of a fair coin is there the highest probability of getting *exactly* half heads and half tails? __100,000__   −2

10. You can toss a coin 2 times, 200 times, 20,000 times or 2,000,000 times. Which option is most likely to result in between 45% and 55% of the tosses landing heads? __20,000 times__

Class Activity 7.2: Simulation

It was not hard to find the probability of flipping heads twice in a row or rolling a 6 and then a 5 with a die. Some probabilities, however, are more difficult to calculate. In those situations, one option is to use a **simulation**, or **Monte Carlo Simulation**. A simulation imitates some chance scenario many times and then approximates an unknown probability. This approximation is justified by the Law of Large Numbers.

Mrs. Qian's advisory makes the following pledge for the annual Faculty-Student Basketball Game: $100 if exactly 4 of the first 5 field goal attempts go in. People on campus hear of this pledge and start wondering how likely it is that Mrs. Qian will have to pay up. Let's assume that there will be at least five field goal attempts in the game, that each field goal has a 40% chance of going in and that all shots are independent of each other. Given this information, calculating the probability that Mrs. Qian has to pay up is relatively easy if you are familiar with the binomial theorem (which all of you will be after we finish chapter 8), but could be a challenging probability to calculate if you are not. For those unfamiliar with the binomial theorem, this would be a good time to try to a simulation in order to estimate the probability that Mrs. Qian has to pay.

The first thing we have to do is to assign digits to represent outcomes. Since 40% of the time a field goal attempt is good, 40% of the digits should correspond to a made field goal and 60% should correspond to a missed field goal. The easiest way to do this is to generate a random integer between 1 and 5 with 1 and 2 corresponding to a "make" and 3, 4 and 5 corresponding to a "miss." To simulate the first five field goals in the basketball game, we are going to use randInt on our calculator. By entering randInt(1,5,5) on the calculator, we will get five randomly selected integers between 1 and 5 inclusive. If you do this on your calculator, your result should look something like the following (obviously the randomly selected integers are likely to be different).

```
randInt(1,5,5)
      {5 5 1 3 3}
```

In this particular simulation, the first five shots were *miss, miss, make, miss, miss*. Since only one of the first five shots was made, Mrs. Qian would not have to pay up. If you hit ENTER, the calculator will generate another five integers. Do this a total of ten times and record your results in the table on the next page.

Trial #	Made Field Goals	Does Mrs. Q pay?
1	1	No
2	3	No
3	3	No
4	2	No
5	4	Yes
6	2	No
7	2	No
8	2	No
9	2	No
10	4	Yes

To estimate the probability of Mrs. Q paying, divide the number of times Mrs. Q has to pay by the number of trials (10).

What is your estimated probability that Mrs. Q has to pay? __20%__

Homework Assignment 7.2: Running a Simulation on Excel

We did a simulation in class but in order for the Law of Large Numbers to work we have to use actual large numbers, and not just ten trials. That is why we should have very little confidence that the estimated probability from Class Activity 7.2 is close to the real probability of Mrs. Q paying. For large and more elaborate simulations, it is easier to use programs like Microsoft® Excel. We are now going to repeat the simulation using Excel

1) On your personal or a school computer, open Microsoft® Excel. Create a new workbook and save it as MrsQPledge.xlsx.

2) We will start off by entering some labels into row 1. In cell A1, type "Trial." In cell B1 through F1, type "First Shot," "Second Shot," "Third Shot," "Fourth Shot," and "Fifth Shot" respectively. In G1, type "Made Field Goals" and finally, "Number of Trials Where Mrs. Q pays" in H1.

3) In A2 through A1001, enter the numbers 1 through 1000. To do this, type the first few numbers, i.e. 1, 2, 3, select these numbers and then drag down the selection from the bottom right. Excel will recognize and continue the arithmetic sequence.

4) In B2, type in the function "=IF(RAND()<0.4, 1, 0)" to have Excel select a number at random between 0 and 1 and return 1 if the number is less than .4 and 0 if the number is not less than .4. This is another way of simulating a field goal attempt with a 40% chance of a make.

5) Select cell B2 and double click on the bottom right corner of the cell. This should fill in all of the cells from B3 to B1001 with the same formula.

6) Copy and paste the formula from B2 into C2, D2, E2, and F2 and select and double click the bottom right of those cells to fill in C3 to F1001.

7) In G2, type in the formula "=SUM(B2:F2)" to add up the five cells to the right. This value should be the number of made field goals in the first five attempts in the trial. Double click on the bottom right of the G2 cell when selected to have G3 to G1001 also sum the five cells to their left.

8) In H2, type in the formula "=COUNTIF(G2:G1001,4)" to count the number of times, in the 1000 trials, that exactly 4 field goals were made.

When you are done, your Excel spreadsheet should look something like the following. Obviously, your random values are likely to be different.

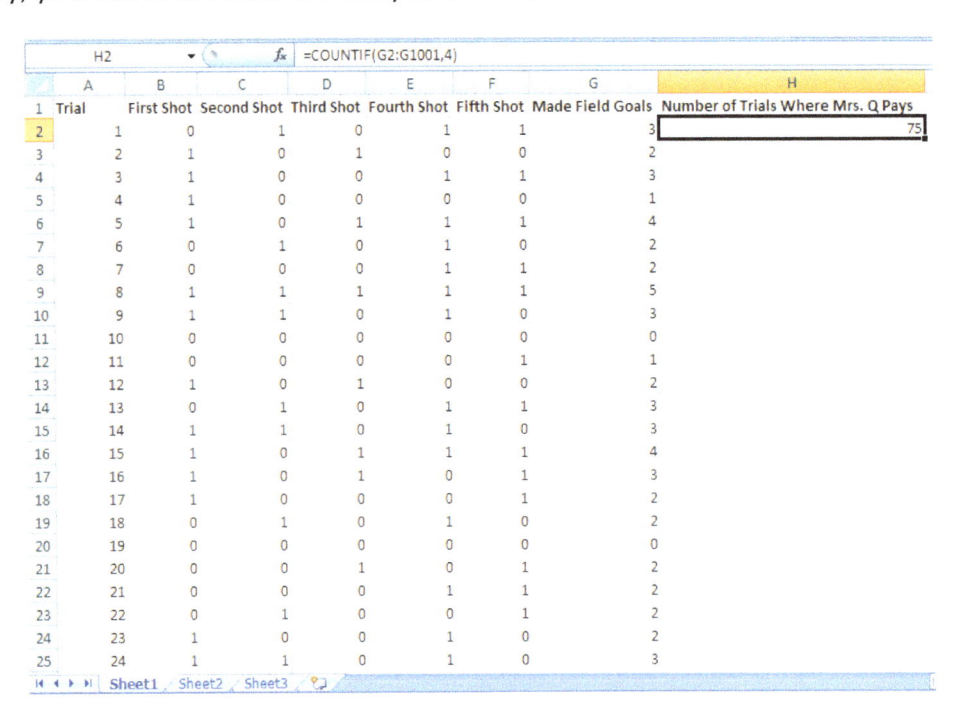

Trial	First Shot	Second Shot	Third Shot	Fourth Shot	Fifth Shot	Made Field Goals	Number of Trials Where Mrs. Q Pays
1	0	1	0	1	1	3	75
2	1	0	1	0	0	2	
3	1	0	0	1	1	3	
4	1	0	0	0	0	1	
5	1	0	1	1	1	4	
6	0	1	0	1	0	2	
7	0	0	0	1	1	2	
8	1	1	1	1	1	5	
9	1	1	0	1	0	3	
10	0	0	0	0	0	0	
11	0	0	0	0	1	1	
12	1	0	1	0	0	2	
13	0	1	0	1	1	3	
14	1	1	0	1	0	3	
15	1	0	1	1	1	4	
16	1	0	1	0	1	3	
17	1	0	0	0	1	2	
18	0	1	0	1	0	2	
19	0	0	0	0	0	0	
20	0	0	1	0	1	2	
21	0	0	0	1	1	2	
22	0	1	0	0	1	2	
23	1	0	0	1	0	2	
24	1	1	0	1	0	3	

9) Save your document and email your teacher the Excel workbook as an attachment. Divide your H2 cell value by 1000 to get your new probability estimate. Put this estimate below.

Estimated Probability = _____

Class Activity 7.3: Probability Models

Simulations are very useful tools when trying to estimate probabilities but it is good to, whenever possible, calculate exact probabilities. For simple random phenomena, a **probability model** is an effective tool for organizing information. A probability model lists all possible outcomes (this list is called the **sample space**) and the probability of each outcome. Often times we are interested in the probability of an **event**, or collection of outcomes. If two events have no outcomes in common we call those events **mutually exclusive**, or disjoint. As mentioned before, when knowing if one event happened does not change the probability of another event happening, those two events are said to be independent.

For example, we might wonder what the probability is of flipping two tails when we toss a coin three times. Let's call this event A. This could happen by flipping T-T-H, T-H-T, or H-T-T. Or, we might wonder the likelihood of flipping three tails when we toss a coin three times. Let's call this event B. This could only happen if the coins come up T-T-T. Or, we might wonder the likelihood of having the first flip land heads when we toss a coin three times. Let's call this event C. This could happen if the coins come up H-T-T, H-H-T, H-T-H, or H-H-H. Finally, we might wonder the likelihood of having the second flip land heads when we toss a coin three times. Let's call this event D. This could happen if the coins come up H-H-T, H-H-H, T-H-T, or T-H-H.

P(A) is how we write "the probability of A." What if we want to know the probability that some event *does not* occur? For example, what if we wanted to know the probability that two tails do not come up when we toss a coin three times? $P(A^c)$ is how we write "the probability of not A." The superscript c stands for complement. What if we want to know the probability that one event *or another* occurs? For example, what if we wanted to know the probability that two *or* three tails come up when we toss a coin three times? $P(A \cup B)$ is how we write "the probability of A or B." $P(A \cup B)$ is the probability that A occurs, B occurs, or that both A and B occur. This is called the *union* of A and B. What if we want to know the probability that one event *and another* occurs? For example, what if we wanted to know the probability that two tails come up *and* that the first flip lands heads when we toss a coin three times? $P(A \cap C)$ is how we write "the probability of A and C." This is called the *intersection* of A and B. Lastly, what if we want to know the probability that one event occurs *given some information*? For example, what if we wanted to know the probability that two tails come up *given* that the first flip land heads when we toss a coin three times? $P(A|C)$ is how we write "the probability of A given C."

Class Activity 7.4: Two-Way Tables

A **two-way table** is a chart used to display the sample space of a chance process when we are considering two events. For example, let's fill in the two-way table below which considers the chance process of drawing a single card from a well-shuffled deck when we are concerned with two events, drawing a spade and getting a face card.

$\frac{12}{52} = $ face

$\frac{13}{52} = $ spade

	B Spade	Not a Spade	Total
A Face Card	3/52	9/52	27/2704
Not a Face Card	10/52	30/52	75/676
Total	15/1352	135/1352	

$\frac{3}{52} = $ spade & face

Homework Assignment 7.3: Using a Two-Way Table

If we use the table above and define drawing a face card as A and drawing a spade as B, find the following probabilities.

P(A) = 12/52   P(B) = 13/52

P(A^c) = 40/52   P(B^c) = 39/52

P(A∪B) = 273/884   P(A∩B) = 1/17

P(A|B) = 15/442   P(B|A) = 15/442

$\frac{9}{52} \cdot \frac{10}{51}$ $\frac{12}{52}$

134

Class Activity 7.5: Venn Diagrams

A **Venn diagram**, much like a two-way table, can illustrate the possible outcomes of a chance process. The most common Venn diagrams consider two or three events, but there are also Venn diagrams that consider the outcomes when four or five events are considered. To understand how Venn diagrams work, let's fill in the number of outcomes that correspond to the four regions in the Venn diagram below: face card and not spade, face card and spade, not face card and spade, and not face card and not spade.

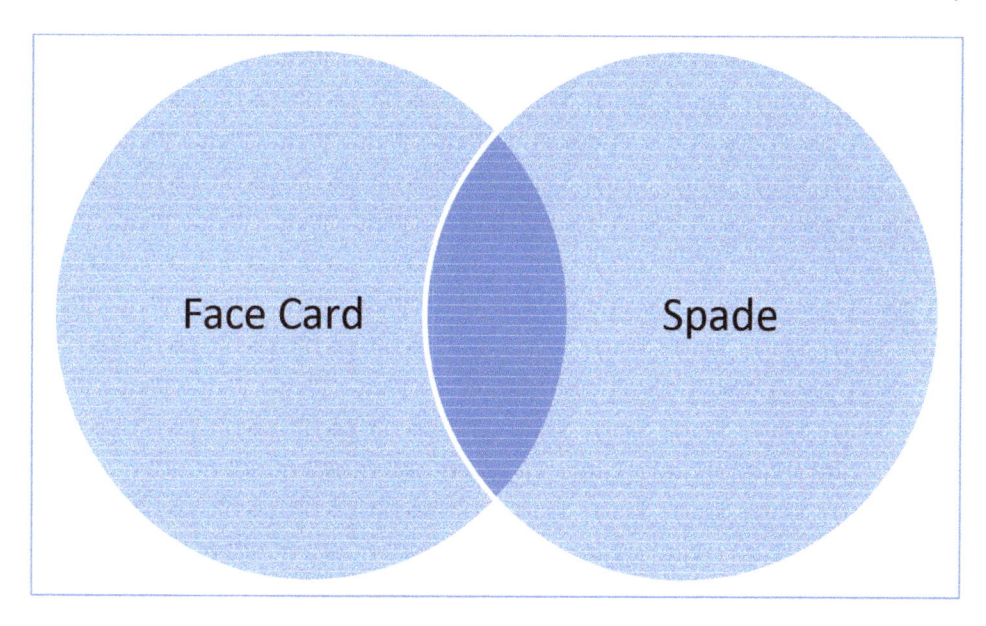

What would a Venn diagram look like for two disjoint events?
_____

Let's fill in the following two-way table for our class considering whether or not a student is 18 and whether or not he has been to Washington D.C.

	18	Not 18	Total
Been to DC			
Not Been to DC			
Total			

135

Homework Assignment 7.4: Making and Interpreting a Venn Diagram

Using the two-way table we made in Class Activity 7.5, fill in the Venn Diagram below with the number of students in each of the four groups.

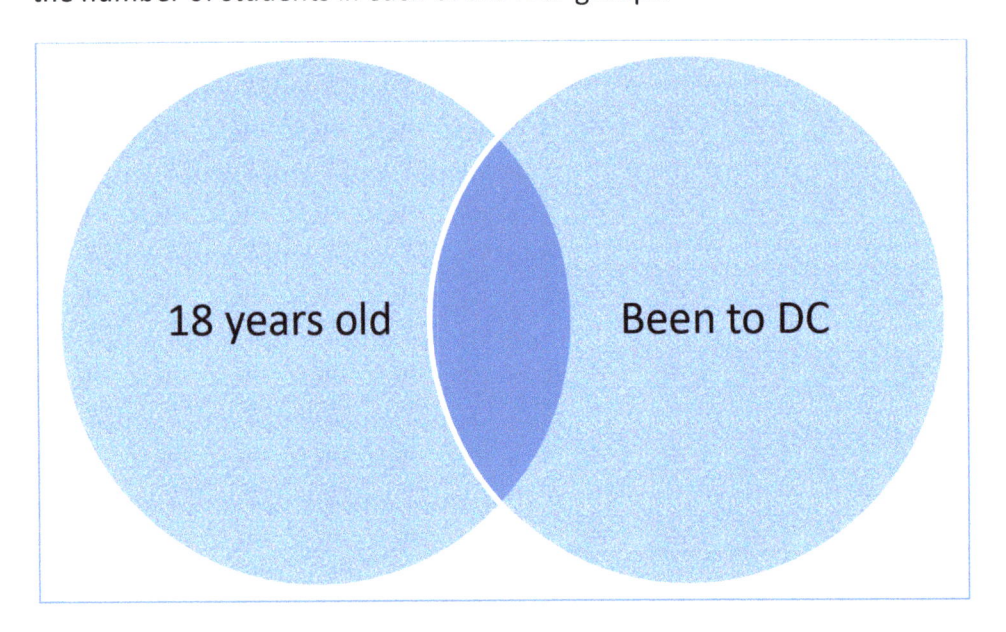

(Venn Diagram with two overlapping circles labeled "18 years old" and "Been to DC")

Using E = is 18 and D = has been to DC, find the following probabilities.

P(E) = 12/52   P(D) = 13/52

P(E^c) = 40/52   P(D^c) = 39/52

P(E∪D) = 313/884   P(E∩D) = 1/17

P(E|D) = 15/442   P(D|E) = 15/442

Class Activity 7.6: Conditional Probability and Tree Diagrams

Throughout this chapter, you have found a number of conditional probabilities, that is, probabilities given certain information. In fact, the last two problems on Homework Assignment 7.4 were conditional probability questions. To make sure you understand conditional probability fully before we move on to tree diagrams, imagine that you are selecting at random one of the approximately 7.5 billion people on Earth. Let W = weighs 300 or more pounds and S = is a professional sumo wrestler. Put the following probabilities in order *from least to greatest*: P(W), P(S), P(W|S) and P(S|W).

_____ , _____ , _____ , _____

A **tree diagram** models chance behavior when there is a sequence of outcomes, i.e. when there are multiple steps involved in the chance process like tossing a coin repeatedly or drawing some cards from a deck. Let's consider the following tree diagram which would be relevant when considering pocket cards in Texas Hold 'Em. There are six branches on the tree diagram, let's fill in the probability that corresponds to each.

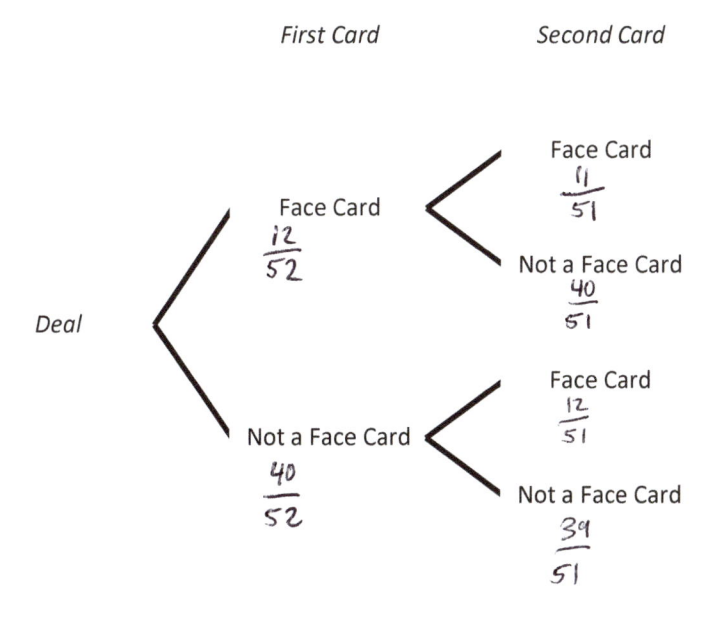

First Card   *Second Card*

Deal
- Face Card $\frac{12}{52}$
  - Face Card $\frac{11}{51}$
  - Not a Face Card $\frac{40}{51}$
- Not a Face Card $\frac{40}{52}$
  - Face Card $\frac{12}{51}$
  - Not a Face Card $\frac{39}{51}$

Hopefully you noticed that getting a face card for your second card is not independent of getting a face card for your first card, i.e. if you get a face card for your first card it is less likely that you will get one for your second card and if you get a card that is not a face card you are more likely to get a face card for your second card. What would have to change about this situation for these two events to become independent?

_____

_____

If two events, A and B, are independent, the probability of both A and B occurring is simply the product of the two individual probabilities, i.e. P(A∩B) = P(A) · P(B) if A and B are independent. If the two events are not independent, however, you need to multiply the probability of A by the probability of B *given A*, i.e. P(A∩B) = P(A) · P(B|A). By dividing both sides of this equation by P(A), you also quickly derive the formula for conditional probability as

$$P(B|A) = P(A \cap B)/P(A).$$

Homework Assignment 7.5: Interpreting Tree Diagrams

Answer the following questions using the tree diagram in Class Activity 7.6.

What is the probability of being dealt two face cards for your pocket cards?

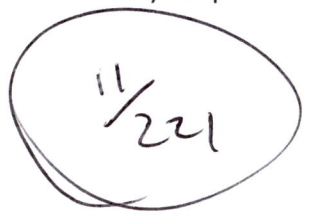

$11/221$

What is the probability of NOT being dealt two face cards for your pocket cards? Find this answer two different ways.

$10/17$

What is the probability of being dealt one face card and one non face card?

$80/221$

Class Activity 7.7: Determining Our General Probability Rules

Now that we have been introduced to most of the basics of probability, let's compile what we have learned with the following set of rules.

Probability Rules

1) If all outcomes in a sample space are equally likely, then

$$P(A) = \frac{\text{number of outcomes corresponding to event } A}{\text{total number of outcomes in the sample space}}$$

2) For any event A:

$\underline{\quad 0 \quad} \leq P(A) \leq \underline{\quad 1 \quad}$

3) If S is the entire sample space:

$P(S) = \underline{\quad 1 \quad}$

4) For any event A:

$P(A^c) = \underline{1 - P(A)}$

5) In general:

$P(A \cup B) = \underline{P(A) + P(B) - P(A \cap B)}$

If A and B are disjoint events then: (mutually exclusive)

$P(A \cup B) = \underline{P(A) + P(B)}$ because $\underline{P(A \cap B) = 0}$

6) If A and B are independent events then:

$P(A|B) = \underline{\frac{P(A \cap B)}{P(B)}}$, $P(B|A) = \underline{\frac{P(A \cap B)}{P(A)}}$ and

$P(A \cap B) = \underline{P(A) \cdot P(B)}$

In general, if A and B are any two events

$P(A \cap B) = \underline{P(A) \cdot P(B|A)}$ which means that

$P(B|A) = \dfrac{P(A \cap B)}{P(A)}$

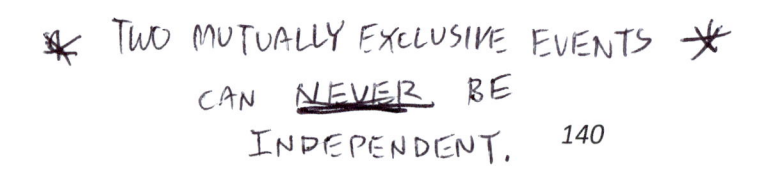

★ **TWO MUTUALLY EXCLUSIVE EVENTS** ★
  CAN <u>NEVER</u> BE
    INDEPENDENT.

Class Activity 7.8: Using Our General Probability Rules

Now that we have listed our basic probability formulas, let's do some problems to make sure we are comfortable using them. The number of the problem corresponds to the number of the rule we are using.

1) Mr. Davidson is playing Monopoly and is on Pacific Avenue. He needs to avoid rolling a 6 or an 8 because his opponent has hotels on Park Place and Boardwalk. What is the probability that Mr. Davidson rolls a 6 or an 8 when he rolls two dice?

2) Which of the following is not a possible probability for some event A?

    0   2/5   .99   4/3   .1

3) When a die is tossed, what is the probability of the result being some number between 1 and 6 inclusive? _____

4) The probability of being dealt five cards from a well-shuffled deck and getting a Royal Flush is 1/649,740. What is the probability of being dealt five cards from a well-shuffled deck and NOT getting a Royal Flush? _____

5) When you draw a card from a well-shuffled deck, what is the probability that you will draw a club or a 7 (Note: this is an inclusive "or" which means that the card could be a club, a 7, or both and satisfy this condition)?

What is the probability that you will draw a club or a red card?

6) What is the probability of being dealt Pocket Rockets (two aces) as your pocket cards in Texas Hold 'Em?

Homework Assignment 7.6: Introducing Simpson's Paradox

Consider two MBA baseball players. They both started playing varsity as sophomores. In the table below you can see the hits, at bats and batting averages for sophomore through senior year.

	Sophomore		Junior		Senior		Combined	
Albert	2/7	.286	18/50	.360	36/105	.343	56/162	0.436
Bob	18/57	.316	22/61	.361	14/39	.359	54/157	0.344

Add up Albert's and Bob's hits and at bats and divide to find their overall batting average to three significant digits.

Albert → 0.286 + 0.360 + 0.343 = 0.989/3 = **0.330**

Bob → 0.316 + 0.361 + 0.359 = 1.036/3 = **0.345**

Who had the better batting average each of the three years? **Bob**

Who had the better overall batting average? **Albert**

How is this possible? *Albert's batting average is overall better because in his sophomore year, he only batted seven times. The worse average for his sophomore year therefore doesn't weigh as much as his average from his senior year (with 105 at-bats) toward his combined batting average.*

Simpson's Paradox / Yule-Simpson Effect

➢ A Paradox in statistics in which a trend appears in different groups of data, but dissapears or reverses when these groups are combined.

➢ Also known as the "Reversal Paradox" or the "Amalgamation Paradox".

➢ It is a special case of what is called ommitted variable bias

Class Activity 7.9: Drug Testing at MBA and Conditional Probability

No drug test, including MBA's drug test, is perfect. When a drug test says that someone used drugs even though they did not, that is known as a **false positive**. When a drug test says that someone did not do drugs even though they did, that is known as a **false negative**.

Let's say that over the next 100 years, MBA will drug test 20,000 students. Let's say that 1% of MBA students actually do drugs and that none are tested twice. It is quite possible that 1% is highly inaccurate but it was chosen to help illustrate a point. Let's say that the MBA drug test has a 1% false positive rate, meaning 1% of the time a person who does not do drugs takes the test it will say he does, and a .5% false negative rate, meaning that .5% of the time a person who does do drugs takes the test it will say he does not. What is the probability that a person who tests positive actually does drugs? With such low false positive and false negative rates, most people would think there is a very good chance that a person who tests positive actually does drugs. Let's find out.

Method 1: Tree Diagram and Conditional Probability Formula

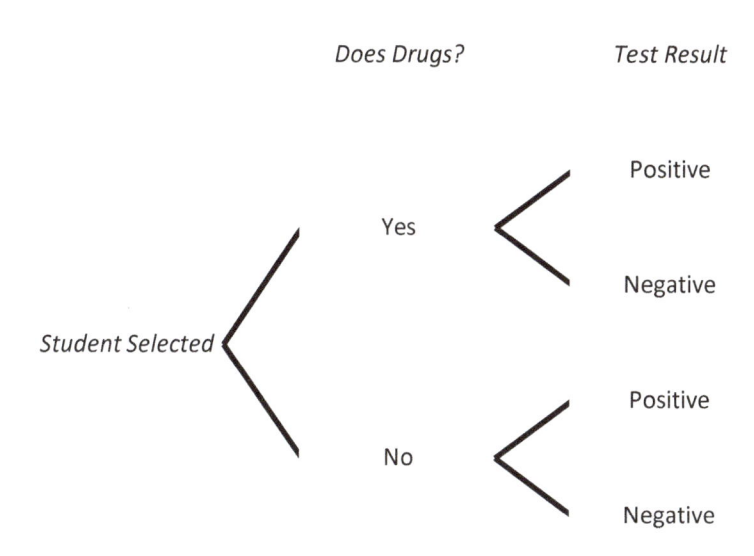

Fill in the appropriate probability on each of the six branches. Let A = positive test result and B = took drugs. Use the conditional probability formula to find P(B|A).

Method 2: Two-Way Table

Fill in the two-way table below and then use the values to find P(B|A).

	Drugs	No Drugs	Total
Positive Result			
Negative Result			
Total			

Class Activity 7.10: 5 Seconds on the Clock...

There are 5 seconds left in a varsity basketball game against Ensworth. MBA has the ball and is down 56-54. Coach Anglin is trying to decide whether to go for a two-pointer in the hopes of forcing overtime or a three-pointer to get the win. Coach Anglin calls a full timeout and then turns to you, the statistics wizard in the crowd, to make the call. Since Coach Anglin is a math teacher, he will only be satisfied if you justify your answer with tree diagrams. In the space below, draw two tree diagrams, one for attempting a two-pointer and one for attempting a three-pointer, complete with probabilities on every branch and then determine which course of action maximizes the probability that we win the game. At least one of your tree diagrams should have multiple steps. You can either make an educated guess or do some research to determine your probabilities.

Class Activity 7.11: Does the "Hot Hand" Exist?

Anyone who has played pick-up basketball is familiar with the idea of the "hot hand." The idea is that, after making a shot or two, a player becomes "hot" and is more likely to make future shots. In 1985, the psychologists Thomas Gilovich of Cornell University and Robert Vallone and Amos Tversky from Stanford researched players from the 76ers, Celtics and the varsity men's and women's basketball teams from Cornell and argued that there was no such thing as the "hot hand." They claimed that streaks of makes and misses by players in basketball games are just like the streaks of heads and tails when you toss a coin. They did not say that everyone has a 50% chance of making or missing a shot, they acknowledged the existence of skill affecting a player's field goal percentage but they argued that the likelihood of a player making a particular shot is not affected by whether he just made or missed his previous shot.[33] In statistical terms, the psychologists argued that field goals attempts are *independent*. Since that article was written in 1985, hundreds have been written, many in support of the claim and a number in opposition. Today in class, we are going to gather some data to see if we believe in the hot hand. Shoot 41 free throws. In the table below, put a √ if you make the free throw and an X if you miss.

#	√ or X	#	√ or X	#	√ or X	#	√ or X
1	X	11	√	21	X	31	X
2	√	12	√	22	√	32	X
3	X	13	X	23	X	33	X
4	X	14	X	24	√	34	X
5	X	15	√	25	√	35	X
6	X	16	√	26	√	36	X
7	X	17	X	27	X	37	X
8	X	18	X	28	X	38	X
9	√	19	X	29	√	39	√
10	√	20	X	30	X	40	X
						41	X

13

Homework Assignment 7.7: Do You Have the Hot Hand?

How many free throws did you make out of your **first 40** shots? __13__

How many free throws did you miss out of your **first 40** shots? __27__

How many free throws did you *make after a make* for shots 2-41? __5__

How many free throws did you *make after a miss* for shots 2-41? __7__

Find P(make free throw | just made a free throw). __5/13__

Find P(make free throw | just missed a free throw). __7/13__

Were you more likely, less likely, or just as likely to make a free throw if you just made one than if you just missed one?
__less likely__

Do you believe that your results provide compelling data for the existence of the hot hand? If yes, why? If no, why not?
__Absolutely not because of my large sample size that actually indicated the opposite of the hot hand.__

Chapter 8. Modeling with Probability

Class Activity 8.1: Expected Value

On an American roulette wheel, there are 38 numbered slots: 1 through 36 are all red or black and 0 and 00 are green (the European roulette wheel does not have the 00). There are a number of bets a player can make in roulette. Two of the more common bets are betting on one of the colors (red or black) or betting on one of the numbers. If a player successfully guesses the color of the slot that the ball lands in, he wins the amount of money that he bet. If he guesses wrong, he loses that bet. For example, if a player bets a dollar on red and it lands red, the croupier (the person running the table) gives him his dollar back and gives him an additional dollar, if it lands on black or green, he loses his dollar bet. It is less likely that a person will correctly predict the number than the color, so the reward for picking the number correctly is greater. If a player correctly guesses the number of the slot that the ball lands in, that bet pays 35 to 1. This means that if a player bets one dollar, he would get his dollar back and in addition he would get 35 dollars from the croupier. If he was incorrect in his bet, he would lose his dollar bet.

So what can a player expect if he starts playing roulette? Should he expect his bankroll to increase or decrease and how quickly? The same law of large numbers we discussed in chapter 7 comes into play here. Just like we can expect that the proportion of times a particular outcome occurs will approach a single value, its probability, we can also expect the result of some chance process with values assigned to outcomes to approach a single value, its *expected value*. Expected value can thus be thought of as the long-run average of some chance process. In the case of the roulette wheel or any gambling, expected value refers to the amount you can expect to win or lose on average per round of playing a particular game. Expected value is found by summing the products of every possible value and the probability that value occurs, i.e. when possible values are $x_1, x_2..., x_k$ and probabilities are $p_1, p_2..., p_k$:

$$\text{Expected Value} = x_1 \cdot p_1 + x_2 \cdot p_2 + ... + x_k \cdot p_k$$

Let's use this formula to calculate the expected value of a $1 bet on red on an American roulette wheel:

Now let's use this formula to calculate the expected value of a $1 bet on 29 on an American roulette wheel:

You can see from these results that the term expected value can actually be a little misleading. If I made a $1 bet on red on an American roulette wheel I expect, in the normal sense of the word, one of two possible outcomes, that I win a dollar or lose a dollar. You will never place a $1 bet on red in roulette and have the result on a single bet be the expected value. Remember that expected value really refers to what you expect in the long run to be your average result.

Class Activity 8.2: Expected Value with Tree Diagrams

Now that we have seen that the law of large numbers applies even when values are applied to outcomes and that we can predict a long-run average using our expected value formula, let's add a tool we learned in chapter 7, the tree diagram. The tree diagram was used in chapter 7 to find probabilities. Now we can add corresponding values to those outcomes, multiply them by probabilities, and find expected values in the same way that we found the expected values of roulette bets in Class Activity 8.1

Imagine that you're at your 20 year MBA reunion and a classmate comes up to you and says he's going to make you an offer that you won't be able to refuse.[34] He says, "I'm on a team developing a cure for gray hair. If successful, taking one pill a year will prevent gray hair entirely. We do, however, need some investors. If you give me $500,000 and the product gets to market, however, you will get a return on your investment of $10 million ($9,500,000 net)." Should you do it? It's not an easy question. The first problem is that we need to figure out how likely it is that the product gets to market and developing a pill that prevents gray hair is not like rolling a die or flipping a coin where we can assume that all outcomes are equally likely. At best we can consider similar attempts to develop drugs in the past and see how often they were successful at various stages. Let's say that you tell your friend that you will get back to him in a week and you go home to do some research.

You find that groups like your friend's are successful in their attempts to develop cures about 30% of the time. If the group is not successful, your friend has assured you that you will get $100,000 of your $500,000 investment back because 20% of funds are dedicated to taking the drug to market. If the drug works, they are approved by the FDA about 50% of the time. Finally, if the FDA approves the drug, there is about an 80% chance of beating other competitors to market. After developing the cure, if the FDA does not approve the drug or if it is approved but is then beaten to market, you will lose your entire $500,000 investment. If the drug is developed, and approved, and is the first drug to market, you will net $9,500,000.

Draw a tree diagram that shows the three steps of developing the drug, getting FDA approval and beating competitors to market with the appropriate probabilities on each branch of the tree below:

Now calculate the expected value of your investment by summing the products of the value (net money gained or lost) that corresponds with each outcome and its probability.

Expected value: _____

So, should you make the investment? Why? If you said no, are there conditions under which you would say yes? If you said yes, are there conditions under which you would say no? What does this exercise tell us about expected value?

_____
_____
_____
_____
_____

Homework Assignment 8.1: When Should You Buy Insurance?

Insurance can be a very complicated business but in very simple terms, the insurance industry makes money by having a positive expected value per customer and thus by collecting more money in premiums than it pays out in claims. To understand this better, let's say your car is insured for $30,000 and the insurance company will pay this $30,000 if the car is stolen or destroyed in an accident or by a natural disaster. The insurance company may determine that the probability of a person like you with a car like yours being stolen or destroyed in the next year is .01.

If that's the case, the annual premium for the car would have to be at least how much money for the insurance company to be making money in the long run by selling policies like this to customers like you? ___*$300.01*___

Of course, the insurance company also needs to pay its bills, employees, etc. so it's likely to charge more than this value. If this is the case, having an insurance policy for you has a negative expected value. The average result from owning an insurance policy is a loss for the customer, i.e. the expected value from the customer's point of view is negative. Then why get auto insurance, health insurance, life insurance etc.? As we saw in the gray hair pill example, decision making is not always as easy as maximizing expected value. People buy insurance to limit the possibility of a disaster. Many people do not have the disposable income to buy a new car if their current car is stolen or destroyed but can pay an insurance premium. If getting a car stolen or destroyed and not being able to replace it would be disastrous (if there would be no other way to get to work, take children to school etc.), it could be well worth the cost of an insurance premium to avoid this possibility, however improbable. In addition, it should be mentioned that in the case of auto insurance, liability car insurance which pays for accidents you cause is typically required by state law.

But what about insurance coverage when the worst case scenario without insurance would not be all that bad? In Charles Wheelan's "Naked Statistics" he titled one chapter "Basic Probability: Don't buy the extended warranty on your $99 printer" and made the argument that when losses would not be catastrophic, the decision making is as simple as maximizing expected value. Which means buying a warranty (which is essentially an insurance policy) on a $99 printer does not make much financial sense unless it would be a total disaster if the printer broke.[35]

What about the super wealthy? In which situations should the super wealthy get insurance (auto, home, health, etc.)? Why?

When it surpasses the cost of paying full for a damaged product as opposed to the insurance, get extended warranty

Class Activity 8.3: Random Variables and a Brown's Diner Burger

A **random variable** takes numerical values corresponding to the outcome of some random process. For example, the *number of heads* that result when tossing a coin 10 times, the *number of 5s* that result when rolling a die twice, and the *number of boys* a mother who has 5 children has would all be random variables.

Let's say that Brown's Diner starts a new promotion. On your 18th birthday, you get to flip a coin at Brown's. If it lands heads, you get a free burger. As it happens, four MBA seniors share a birthday (March 14th). The four seniors decide to head to Brown's for off-campus lunch. When they get to Brown's, they all flip the coin and every single senior gets heads! Brown's is happy that the customers are happy, but wonders how many free burgers they are going to give out this year.

Define H = the number of heads. Let's make a **probability distribution** for this random variable. A probability distribution is like a probability model except instead of simply listing the outcomes we list the numerical values that our variable can take and can appear in table or graph form.

Value	0	1	2	3	4
Probability					

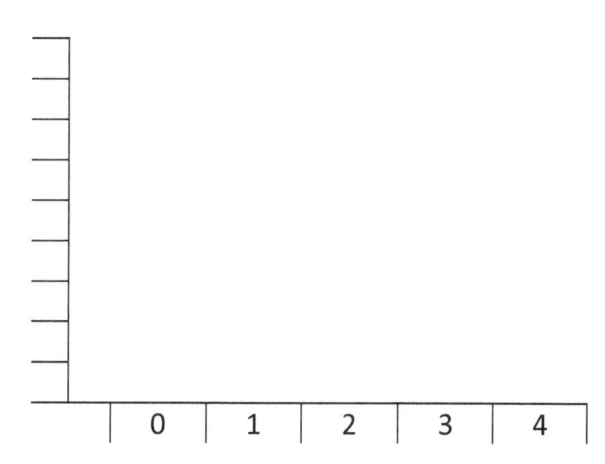

If another group of 4 students comes into Brown's Diner on March 15th all celebrating their birthdays, what is the probability that all 4 will get free burgers? _____

Use our expected value formula to determine the average number of free burgers Brown's would expect to give out per 4 birthdays. _____

For a random variable, when there are *n* independent trials with a probability of success *p*, expected value can be found with a much simpler formula:

Expected Value = np

Using this new formula, determine the average number of free burgers Brown's would expect to give out per 4 birthdays. _____

It is not terribly tedious to list all of the 16 possible outcomes when you toss a coin 4 times and then see which outcomes correspond to the different number of heads. Since all 16 outcomes are equally likely, determining the probability distribution is just a matter of division. But what if octuplets come into Brown's one day? If all eight octuplets flip a coin, the number of possible outcomes is 256. Writing out all 256 outcomes to determine the probability distribution would be very time consuming. One alternative is to use **Pascal's Triangle**. Pascal's Triangle continues into infinity but it starts like the following:

order does not matter

```
                    1
                 1     1
              1     2     1
           1     3     3     1
        1     4     6     4     1
     1     5    10    10     5     1
```

You may remember Pascal's triangle from binomial expansion. For example, $(a + b)^4 = a^4 + 4a^3b + 6a^2b^2 + 4ab^3 + b^4$. The coefficients of that expansion are 1, 4, 6, 4, and 1 which are the values of row 4. You might think this is really row 5 but the top row that consists of a single 1 is referred to as row 0. You'll see why in a second. So what does this have to do with probability? Well you may also recognize that those values in row 4 refer to the number of outcomes corresponding to each numerical value of random variable H when H = number of heads that result when tossing a coin four times. Confused? Think about it this way. The top row refers to the possible outcomes when a coin is tossed 0 times. What is the only possible outcome? 0 heads and 0 tails. What if you toss a coin once? Then you could have either 1 head or 1 tail. That's what the 1 1 refers to in row 1. After two tosses, there are four possible outcomes which correspond to three possible numerical values for H. You could have two heads (from H-H), one head and one tail (from H-T or T-H) or you could have zero heads and two tails (from T-T). The value of a cell in Pascal's Triangle equals the sum of the two values directly above it.

Counting principal = multiply the choices of each category

1. Counting Principle — Order doesn't matter
2. Factorial — used in calculation nPr, nCr
3. Permutation — (math → prob → nPr), order matters: $nP_k = \frac{n!}{(n-k)!}$, $n > k$
4. Combination — (math → prob → nCr), order doesn't matter: $nC_k = \frac{nP_k}{k!} = \frac{n!}{k!(n-k)!}$

Homework Assignment 8.2: Octuplets Walk Into Brown's Diner

Eight octuplets decide to head to Brown's for dinner. When they get to Brown's, they all flip the coin to determine if they are a winner. Define H = the number of heads. Let's make a **probability distribution** for this random variable in table and graph form. You can list all 256 outcomes but I highly recommend drawing and using Pascal's Triangle.

```
                    1
                  1   1
                1   2   1
              1   3   3   1
            1   4   6   4   1
          1   5  10  10   5   1
        1   6  15  10  15   6   1
      1   7  21  35  35  21   7   1
    1   8  28  56  70  56  28   8   1
  1   9  36  84 126 126  84  36   9   1
```

Value	0	1	2	3	4	5	6	7	8
Probability	1	1/2	1/4	1/8	1/16	1/32	1/64	1/128	1/256

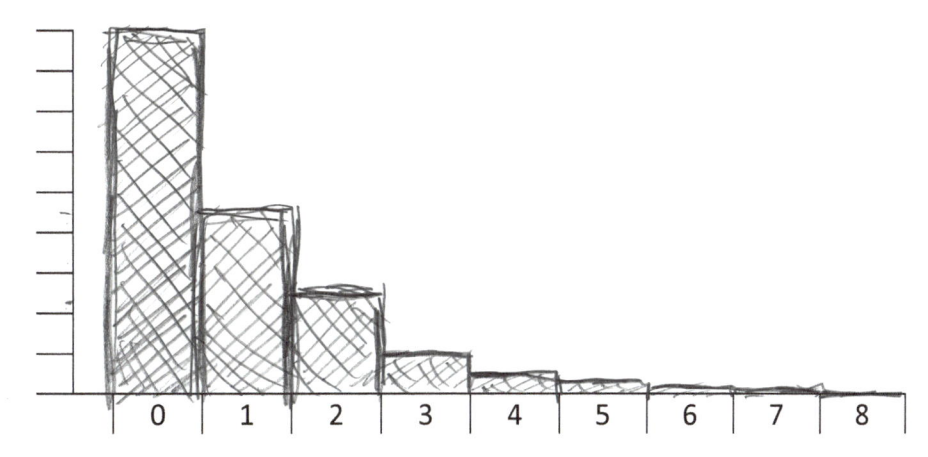

Use our first expected value (Expected Value = $x_1 \cdot p_1 + x_2 \cdot p_2 + \ldots + x_k \cdot p_k$) to determine the average number of free burgers Brown's would expect to give out per octuplet birthday party. Show your work.

$$(0)(1) + (1)\left(\frac{1}{2}\right) + (2)\left(\frac{1}{4}\right) + (3)\left(\frac{1}{8}\right) + (4)\left(\frac{1}{16}\right) + (5)\left(\frac{1}{32}\right) + (6)\left(\frac{1}{64}\right) + (7)\left(\frac{1}{128}\right) + (8)\left(\frac{1}{256}\right)$$

2

Using our second formula (Expected Value = np), determine the average number of free burgers Brown's would expect to give out per octuplet birthday party. Show your work.

$$EV = np = 8 \times \frac{1}{2} = 4$$

_____ 4

Class Activity 8.4: Brown's Catfish Dinner and Counting 101

Brown's promotion is a success and they decide to invest in a large die and change the promotion to "if you roll a three, then your catfish is free." People like the idea and start coming out in bigger and bigger crowds for their chance at a free catfish dinner. Brown's had a pretty good handle on how many burgers they could reasonably expect to lose in an evening but the calculation is more complex now and the diner only has 70 catfish dinners in stock. If they are expecting about 300 people to roll the die this evening, what is the probability that the diner runs out of catfish? Even with Pascal's Triangle, this problem would take hours without additional tools. There are so many ways that 71 or more of the 300 customers could roll a 3. Before answering the question, we need to get better at counting.

Our first formula for counting is so simple that most people do it for years without even recognizing that it could be thought of as a formula. The **multiplication principle** states that when a process contains k steps and there are n_1 ways to do the first step, n_2 ways to do the second step,..., and n_k ways to do the kth step, the total number of ways to complete the process is:

$$n_1 \cdot n_2 \cdot \ldots \cdot n_k$$

If you get up in the morning and have 2 options for sneakers, 3 options for underwear, 2 options for slacks, 1 option for belt, 5 options for t-shirt, and 2 options for jacket, how many distinct outfits can you choose from? ____120____

Let's say Mr. Davidson is grading tests from his 14-person 3rd period Statistics class. We could use the multiplication counting principle to determine how many possible orders there are. Mr. Davidson will have 14 options for the test he grades first. Regardless of which test he grades first, he will have 13 options for the test he grades second. This pattern continues for the third through 14th test and thus the total number of possibilities is $14 \cdot 13 \cdot 12 \cdot \ldots \cdot 3 \cdot 2 \cdot 1 = 87{,}178{,}291{,}200$. Expressions in the form $n(n-1)(n-2) \cdot \ldots \cdot 3 \cdot 2 \cdot 1$ occur so frequently enough that there is a special name and notation for them. We write

14·13·12·…·3·2·1 as 14! and read it as "14 **factorial**." The factorial symbol "!" can be found on your TI calculator by pressing MATH, scrolling over to PRB and then selecting 4: !. If you find factorials like 14! on your calculator, they are likely to be expressed in scientific notation like the following because the values so quickly become so large.

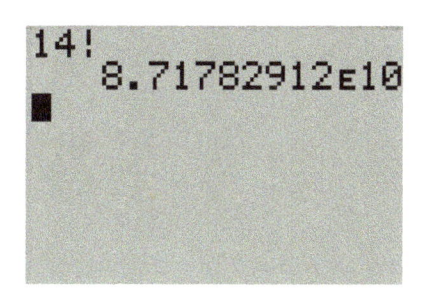

```
14!
       8.71782912E10
```

Homework Assignment 8.3: Multiplication Principle and Factorials

During Nashville Originals Restaurant Week, eateries in Nashville draw up limited menus for discounted prices to entice new customers to dine with them. At one restaurant, customers could choose between one of three appetizers, one of four entrees, and one of two desserts. How many different orders were possible from someone ordering off this special menu? __24__

If a waiter took an order and then completely forgot what the person wanted and told the kitchen to select an appetizer, entrée and dessert at random and hope they got the order right, what is the probability that they would? __1/24__

All the MBA busses are being used so Mr. Davidson drives five students to a special exhibit of "The Beauty of Pi, Phi, and e" at the Frist Center for the Visual Arts. If there are five passenger seats in Mr. D's minivan, in how many different arrangements could the students sit? __120__

At the last minute, one of the students backs out so now there will be four students traveling with Mr. Davidson. How many arrangements of four students in five passenger seats are there? __240__

According to the Tennessee Department of Revenue: Vehicle Services Division, personalized passenger license plates can be between 3 and 7 characters and can contain the letters A through Z and the digits 0 through 9.[36] How many personalized passenger license plate possibilities are there?

$$(36)^7 + (36)^6 + (36)^5 + (36)^4 + (36)^3$$

__8.06×10^{10} license plates__

Class Activity 8.5: Permutations and Combinations in Brentwood

A group of friends from MBA heads to Brentwood Academy to watch the MBA-BA basketball game. While sitting in the crowd, the friends do what friends normally do on a school night, talk about the number of possible line-ups MBA could put on the floor. There are 12 players on MBA's varsity roster and five players can be playing at any particular time in the game. Most teams have one player as point guard, one as a shooting guard, one as a small forward, one as a power forward and one as a center. The different positions have different roles in the game, e.g. the point guard is normally responsible for getting the ball down the court and running the offense.

Xavier, fresh off an eye-opening Statistics class, says that you can find the total number of possible line-ups using the multiplication principle. Say Coach Anglin chooses his center first. He has 12 options. If he chooses his power forward second, he has 11 options. If he chooses his small forward third, he has 10 options, etc. This means that there are 12·11·10·9·8 = 95,040 possible line-ups. Much like factorials, these cases where order of selection matters and you lose one option each step along the way have a special name, **permutations**. 12·11·10·9·8 can be written as $_{12}P_5$ where P stands for permutation. You can find permutations on your calculator. First type the number of individuals in the group (12 in this case), then press MATH, scroll over for PRB, and then select 2: nPr. Finally, type in the number of individuals chosen from the group (5 in this case). If done correctly, when you hit ENTER your screen should look like the following:

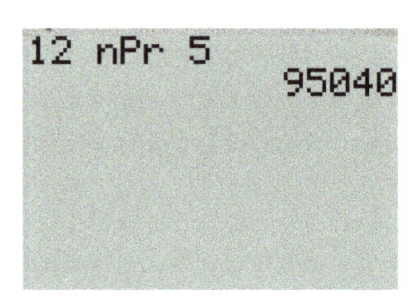

```
12 nPr 5
              95040
```

After an exciting overtime victory, the group of friends decides to head to Blaze Pizza in Brentwood for a celebratory dinner. Yuri had specific instructions from his mother. He had to get classic crust, red sauce, and shredded mozzarella and he was not allowed to get any finishes but his mom said he could choose any 5 toppings he wanted. There are 26 topping options at Blaze Pizza.[37]

Yuri started wondering how many different options he has given his mother's constraints. He thinks he might be able to use the same method that Xavier used to figure out the number of possible line-ups at the basketball game. Then a lightblub goes off! These two situations are different. If you get a pizza with bacon, pepperoni, artichokes, basil and spinach, it is not distinguishable from a pizza with basil, spinach, pepperoni, bacon and artichokes. Both pizzas have the same five ingredients and it would not be possible to say in which order the ingredients were requested when the pizza came out of the oven.

With the basketball team, the order did matter. A basketball team with the 6'11" junior playing point guard is very different from the basketball team with the 6'11" junior playing center. How many possible pizzas can Yuri order then? Well, we can start by finding out how many pizzas are possible assuming that order does matter. Then, we can find out how many of these resulting pizzas essentially are the same as other pizzas. That is, we can figure out how many different ways we could get a pizza with bacon, pepperoni, artichokes, basil and spinach pizza. It just so happens that there are 5! = 120 ways. By dividing the permutation by 5! we get a new number. It is the number of ways of selecting individuals from a group where order does not matter. These types of numbers are called **combinations**. The number of distinct pizzas that Yuri could order can be written as $_{26}C_5$ and it equals $\frac{26 \cdot 25 \cdot 24 \cdot 23 \cdot 22}{5 \cdot 4 \cdot 3 \cdot 2 \cdot 1}$ = 65,780.

You can find combinations on your calculator as well. First type the number of individuals in the group (26 in this case), then press MATH, scroll over for PRB, and then select 3: nCr. Finally, type in the number of individuals chosen from the group (5 in this case). If done correctly, when you hit ENTER your screen should look like the following:

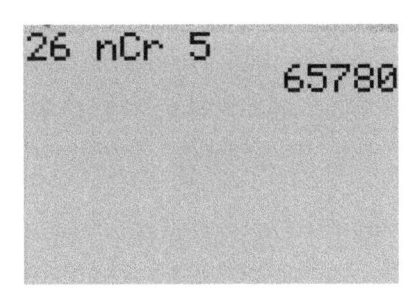

```
26 nCr 5
              65780
```

KNOW "BINS"

Homework Assignment 8.4: Mr. D's Ice Cream Shoppe and Poker Probabilities Revisited

Mr. Davidson opens up an Ice Cream Shoppe in the summer to give himself a chance to stay sharp with combinations and permutations. If he has 8 flavor options, how many ways can Mr. Davidson make:

A cone with four scoops if you care about the order of flavors? **1680**

$$_8P_4 = 8 \cdot 7 \cdot 6 \cdot 5 = 1680$$

A cone with two scoops of *different* flavors if you care about the order of flavors? **56**

$$_8P_2 = 8 \cdot 7 = 56$$

A cup with three scoops of *different* flavors (order doesn't matter)? **56**

$$_8C_3$$

A cup with three scoops of ice cream (order doesn't matter)? **70**

$$_8C_4$$

His famous "Davidson Skyscraper" cone with all 8 scoops of *different* flavors if you care about the order of flavors? **40320**

$$_8P_8$$

His less famous "Davidson Building" cone with 7 scoops of *different* flavors if you care about the order of flavors? **5040**

$$7!$$

How many possible pocket cards can be dealt from a 52-card deck? Remember that pocket cards are two cards and that the order of the two cards does not matter, i.e. a four of diamonds and a three of clubs is the same hand as a three of clubs and four of diamonds. **1326**

How many possible pocket cards can be dealt from a 52-card deck that contain two face cards? **66**

$$_{12}C_2$$

If all pocket card combinations are equally likely, what is the probability of being dealt pocket cards that are both face cards? **0.050**

$$\frac{12}{52} \cdot \frac{11}{51} = \frac{11}{221}$$

Compare this answer to your answer from Homework Assignment 7.5

$$\approx 0.050$$

Class Activity 8.6: The Binomial Theorem at the UT-Vandy Game

Now that we have introduced combinations, we can talk about the binomial theorem and then hopefully we get back to helping Brown's Diner with its catfish debacle.

Tennessee is playing Vanderbilt in basketball. Vanderbilt is down 74-71 with 10 seconds left. Vanderbilt's point guard passes to the shooting guard who pump fakes, gets the defender in the air, and then gets the shot off right before the buzzer sounds. The ball falls short but, a whistle!, the referee says that the defender fouled the shooter and the player should be awarded three free throws. The Tennessee coach cannot believe it and storms onto the court screaming. The coach is awarded a Class A technical foul which means that Vanderbilt gets an additional two free throws. With 0 seconds on the clock, Vanderbilt's shooting guard steps up to the free throw line where he will attempt five free throws. The 19 year old has not attempted a free throw tonight but the SEC network displays his season free throw percentage at the bottom of the screen as 80%. Across the state and country, Volunteer and Commodore fans wonder if their team is going to win or if this game is going to overtime. More mathematically inclined viewers may wonder, what is the probability of these possible events?

Coin tosses and die rolls are universally accepted to be independent events and the probabilities corresponding to those random phenomena are not thought to change. If you are tossing a fair coin your probability of landing heads is .5 and if you roll a die your probability of getting a 2 is 1/6 and it does not matter what happened on your last toss or roll. In the case of free throws, statisticians and sports enthusiasts might argue that a given player is "clutch" or "not clutch" or appeal to something like the hot hand as a reason why the trials (free throws) might not be independent. For our purposes, however, we are going to assume that every time the shooting guard takes a free throw, there is a .8 probability that it will go in.

Let's start with the easiest question: What is the probability that all five free throws are made? There is only one way for that to happen (make-make-make-make-make), which means the probability would be $(.8)(.8)(.8)(.8)(.8) = (.8)^5 = .32768$. Now, let's take a look at the second easiest question: What is the probability that all five free throws are missed? There is only one way for that to happen (miss-miss-miss-miss-miss), which means the probability would be $(.2)(.2)(.2)(.2)(.2) = (.2)^5 = .00032$.

It seems easy enough so far. Let's try another: What is the probability that four of the free throws are made? This time there are five ways to reach this result (make-make-make-make-miss, make-make-make-miss-make, make-make-miss-make-make, make-miss-make-make-make, and miss-make-make-make-make). Each one of these outcomes has a

$(.8)^4(.2)^1 = .08192$ probability of occurring which means that the probability of *any* of those outcomes occurring is $5(.8)^4(.2)^1 = .4096$. A little trickier, but by the same logic, the probability of having one made free throw is $5(.2)^4(.8)^1 = .0064$.

Now it really gets hard. What is the probability that three of the free throws are made? You can try to list all the ways in which 3 free throws are makes and 2 are misses (make-make-make-miss-miss, make-make-miss-make-miss, etc.), but this method is starting to get painful. What we want to find is the number of ways of selecting three individuals from a group of five where order doesn't matter (making your 1st, 2nd, and 4th free throw is really the same thing as making your 2nd, 4th and 1st free throw). What does this sound like? A combination! By multiplying a combination which represents the number of outcomes that correspond with a given event by the probability of one of those outcomes you can calculate the probability of the event. The **binomial theorem** thus says:

$$P(X = k) = {}_nC_k \cdot p^k \cdot (1-p)^{n-k}$$

Where the variable *X* represents the number of successes, *k* represents the number of successes in *n* trials with a probability of success on each trial *p*.

So far we have found four probabilities. If we were to find them with the binomial theorem it would look like the following:

$P(X = 5) = {}_5C_5 \cdot (.8)^5 \cdot (.2)^0 = .32768$

$P(X = 0) = {}_5C_0 \cdot (.8)^0 \cdot (.2)^5 = .00032$

$P(X = 4) = {}_5C_4 \cdot (.8)^4 \cdot (.2)^1 = .4096$

$P(X = 1) = {}_5C_1 \cdot (.8)^1 \cdot (.2)^4 = .0064$

Now let's find the probabilities of two free throws being made and three free throws being made:

$P(X = 2) = {}_5C_2 \cdot (.8)^2 \cdot (.2)^3 = .0512$

$P(X = 3) = {}_5C_3 \cdot (.8)^3 \cdot (.2)^2 = .2048$

You can find these binomial probabilities on your calculator. Press 2nd, VARS (DISTR) and then scroll down to A: binompdf(. Select it and then enter the number of trials, the probability of success, and the number of successes you want to find the probability of occurring. For example, if you wanted to find our final probability, that of the shooting guard hitting three of his free throws, the screen would look like this:

```
binompdf(5,.8,3)
              .2048
```

Another useful command is binomcdf(. It is in the same list of commands as binompdf(but instead of giving the probability of k successes, it gives the probability of k or fewer successes and thus saves you the trouble of summing multiple results. For example, if you wanted to find the probability that Tennessee will win the game, you would want to find the probability that the guard hits two or fewer free throws. Instead of summing the probability of the guard hitting 0, or 1, or 2, you can use binomcdf(and your result looks like the following:

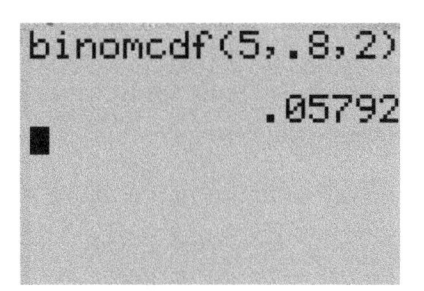

```
binomcdf(5,.8,2)
              .05792
```

You would get the same probability by adding the individual probabilities like thus, P(X = 0) + P(X = 1) + P(X = 2) = .00032 + .0064 + .0512 = .05792.

This binomial theorem is very useful but we have to make sure we are using it appropriately. There is a mnemonic device that can be used to ensure that all four conditions necessary to be a binomial setting are met. Just remember BINS: binary, independent, number and success. To be a **binomial setting**, i.e. a situation where the binomial theorem can be used to calculate probabilities, the first condition is that the outcomes can be classified as a success or failure, i.e. that the outcome is **binary**. Note that even if there are more than two outcomes (like rolling a die and getting some result between 1 and 6) it can still be a binomial setting if you divide the outcomes into successes and failures (1 or 2 is a success and 3, 4, 5 and 6 is a failure for example). Secondly, the trials must be **independent**. If you are picking cards out of a deck looking for clubs with replacement, the trials are independent. If you are picking cards out of a deck without replacement, the trials are dependent and thus it would not be a binomial setting. Thirdly, the **number** of trials must be fixed. Shooting ten free throws qualifies as a binomial setting. Shooting until you hit five free throws would not qualify because it might take five trials and might take fifty. Lastly, the probability of **success** must be the same for each trial.

Homework Assignment 8.5: Is it a Binomial Setting?

Determine whether the following scenarios qualify as binomial settings. If a scenario does, write "binomial setting." If it does not, state which condition is not met.

Draw 6 cards from a deck without replacement and record the number of 3's.

Does not meet rule #2: independence; Not Binomial Setting

Roll a pair of dice over and over until it lands "snake eyes" (both dice show 1). Record the number of rolls it took.

Does not meet rule #3: number; Not Binomial Setting

50% of Tennesseans have brown eyes, 20% have blue eyes, 10% have hazel eyes, and 20% have some other color eyes. Select 20 Tennesseans at random and record whether they have brown, blue, hazel or other colored eyes.

Does not meet rule #1: binary; Not Binomial Setting

Coach McMahan shoots 85% from the free throw line. Coach Anglin shoots 80% from the free throw line. They each shoot 10 free throws and the total number of made free throws out of 20 is recorded. *You could do binomial settings for each shooter individually. But, the binomial setting won't work using both of them because it violates rule #4: consistency in probability of success. Not Binomial Setting*

Ten MBA friends buy scratch-off lottery tickets. The probability of winning something on each ticket is 1 in 3. After they scratch the tickets, they record the number who won something. *Binomial Setting*

KNOW "BINS" FOR BINOMIAL SETTING

Handwritten at top:
B I N S
↑ ↑ ↑ ↑ success
binary independent number

Class Activity 8.7: Solving the Catfish Problem

Now we have the tools necessary to help consult Brown's about the free catfish promotion. As a reminder, 300 people are going to go to Brown's tonight to roll a die. Every time it lands on a three, the person who rolled gets a free catfish dinner. The diner has 70 catfish in stock. We know that the expected number of catfish Brown's will give away is EV = np = (300)(1/6) = 50. However, Brown's is worried about the public relations mess they may get into if they run out of catfish.

Luckily, this is a binomial setting. The probability of success (winning a catfish) is 1/6. There are 300 trials. We want to know the probability of 71 or more catfish being won. We could use the binomial theorem to find the probability of 71 catfish being won, then 72, then 73 etc. but that would take too much time. If we use binomcdf(on our calculator it can calculate the probability of 70 or fewer catfish being won and then we can find the probability of 71 or more being won by subtracting the binomcdf(result from 1 since winning 70 or fewer catfish or 71 or more catfish contain all possible outcomes in this binomial setting. First calculate the probability of 70 or fewer catfish being won on your calculator. When your screen looks like the following, press ENTER.

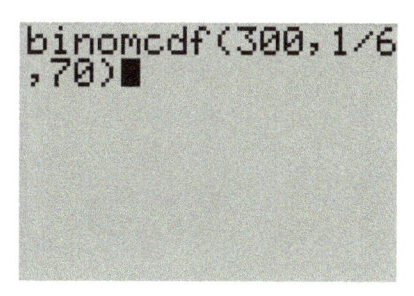

```
binomcdf(300,1/6
,70)
```

What is the probability of 70 or fewer catfish being won? __0.9988__

Now subtract that answer from 1 to find the probability of 71 or more catfish being won. What is the probability that 71 or more catfish are won and that Brown's Diner runs out? __0.00165__

If you were consulting Brown's, what would your recommendation be? Go ahead with the promotion, cancel it, or start looking for more catfish?

Class Activity 8.8: Binomial and Sampling Distributions

I used Excel and the BINOMDIST function to make a histogram of the probability distribution for X = the number of catfish won. Every bar represents the probability of that many catfish being won, e.g. the probability of exactly 56 catfish being won is approximately 0.038818883. The histogram is below:

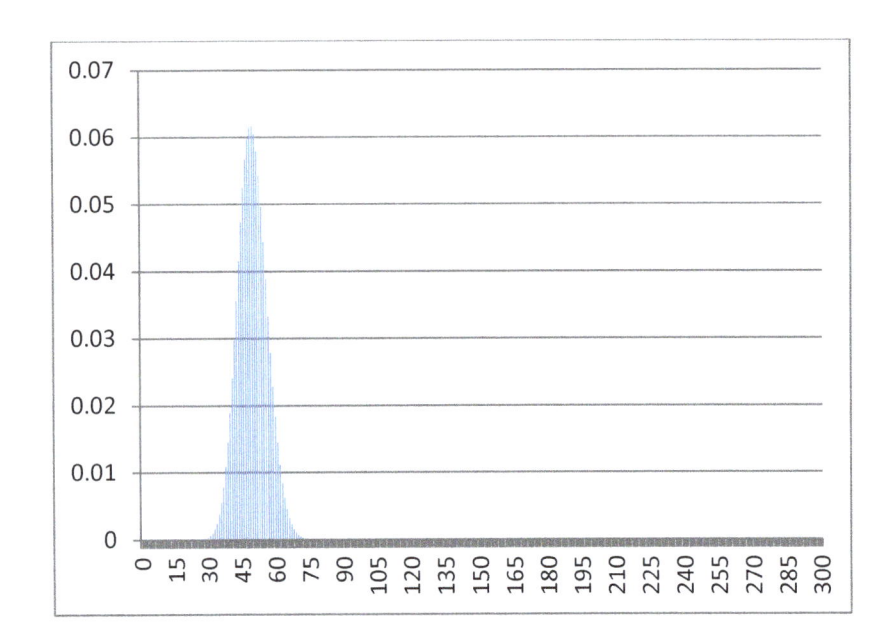

This distribution probably looks a lot like the normal distributions we looked at in chapter 5. As the number of trials *n* gets larger, the binomial distribution gets close to a normal distribution. This means that we can approximate binomial probabilities using the same methods we used on normal distributions in chapter 5.

Let's say that instead of rolling a die to win a catfish at Brown's, people in Davidson County are responding to a survey saying whom they would most like to see elected president in 2016. Let's say that it just so happens that exactly 1/6 of Davidson County would like to see Bernie Sanders as our next Commander in Chief. If 300 people are selected at random for the sample, how likely would it be that 58 say they want to see Bernie win? How about 41? The question ends up being just like the question: How likely would it be that 58 or 41 catfish are won? Using proportions instead of counts, the **sampling**

distribution for this Bernie poll, a representation of possible statistic values and their corresponding probabilities, would look like the following.

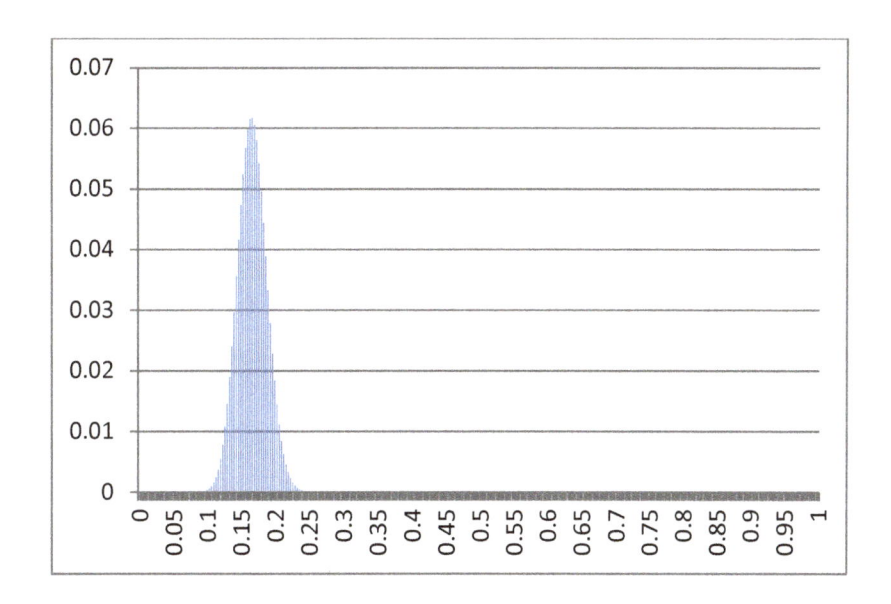

Since the number of catfish winners distribution and proportion of people in poll supporting Bernie distribution are both approximately normal, it is possible to use standard normal probability values and our 68-95-99.7 rule to explore the distributions' characteristics.

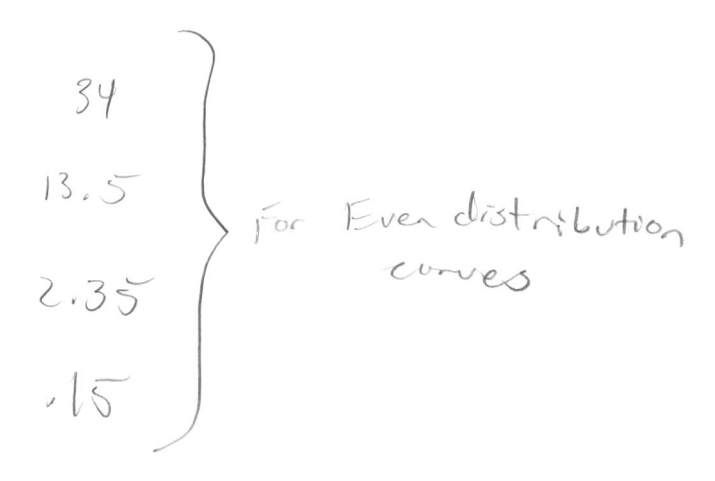

34
13.5
2.35
.15
} For Even distribution curves

Homework Assignment 8.6: Working With Sampling Distributions

An opinion poll asked an SRS of 1600 adults in Tennessee, "Have you read at least one Harry Potter book?" Let's suppose that the population proportion of Tennessee who has is p = .2. If this opinion poll were repeated many times, the statistic $\hat{p}$ would be approximately normally distributed with mean .2 and standard deviation .01. Sketch the normal curve of this sampling distribution below.

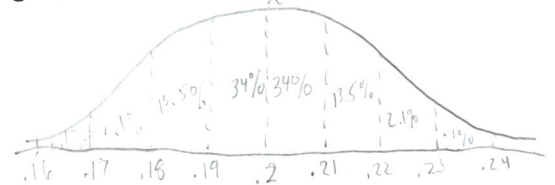

[Hand-drawn normal curve with labels: 2.5%, 13.5%, 34%, 34%, 13.5%, 2.5%, .15% along the x-axis values .16, .17, .18, .19, .2, .21, .22, .23, .24]

Using the 68-95-99.7 Rule, find the probability that between 18% and 22% of those polled say they have read at least one Harry Potter book.

$$P = 0.95$$

Using the 68-95-99.7 Rule, find the probability that between 17% and 21% of those polled say they have read at least one Harry Potter book.

$$P = 0.838$$

We have claimed that taking a simple random sample is a binomial setting. This is very close to true but in reality, trials are not independent, e.g. if 1/6 of the population supports Bernie and the first person you ask says he supports Bernie, the second person you poll is *just barely* less than 1/6 likely to be a Bernie supporter. Strictly speaking, this means this is a hypergeometric distribution and not a binomial one. As long as you are sampling from a large population, the trials are close enough to independent that this doesn't make a difference. But what if your sample made up a large portion of the population? For example, let's say that 1/6 of MBA high school students support Bernie Sanders. If you poll 300 of approximately 500 high school students, how would your sampling distribution be different than the sample discussed in Class Activity 8.7?

$$P = 0.16$$

Population perameter for catfish was 300, but the value 300 in the Bernie sample was not indicative of population perameter; rather it was indicative of sample size.

More uniform curve, less peak in curve

Class Activity 8.9: The Boy Named Banjo Concert

You and a bunch of your classmates decide to spend your senior year college spring break taking a Winnebago to Las Vegas to see Boy Named Banjo play at Caesars Palace. The Winnebago has enough gas to make the round trip and is stocked with enough food to keep you and your friends fed there and back. In addition, every person brought $60 to buy the tickets to the concert when you arrive. Sadly, when the group arrives in Vegas, they find that the $60 tickets have all been sold and that all that remains are $100 tickets. The $60 was all the students could each scrap together for the concert and the concert starts in 4 hours. Then one student has an idea, what if we gamble and try to win $40 each?![38]

Generally, gambling is a great way to lose money. Unless you are a card counting wizard or are playing poker with bad poker players, you're likely to end up losing more than you win. It's not always the case, however, and while the best financial advice you could likely have in Vegas is to keep your money in your pocket, it is a fun application of probability to decide, in a bizarre situation like this, what course of action would give the best chance of turning that $60 into $100 and getting to go to the concert.

In the first section of this chapter, we calculated the expected value of a $1 bet on an American roulette wheel. We saw that you can expect to lose a little over 5 cents per $1 bet in the long run. There are many gambling games but the best expected value (even though it's still negative) is often found at the craps table. Craps is a little more complicated than roulette. Like roulette, there are many options for bets but the most common bet is a pass line bet. If you make a pass line bet, two dice are thrown. If the result is 7 or 11, you win the amount you bet. If the result is 2, 3 or 12, you lose the amount you bet. If any other number results (4, 5, 6, 8, 9, or 10), you continue to roll the dice until the first number rolled comes up again or a 7 comes up. If a 7 comes up, you lose your bet. If the first number rolled comes up again, you win the amount you bet at the start. As you no doubt can see, calculating the expected value of a pass line bet is a lot more complicated than calculating the expected value of a bet on red in roulette. I have compiled the relevant probabilities necessary for calculating the expected value in the table on the next page:

Opening Roll	P(Opening Roll)	P(Win|Opening Roll)	P(Opening Roll ∩ Win)
2	1/36	0	0
3	2/36	0	0
4	3/36	3/9	.0278
5	4/36	4/10	.0444
6	5/36	5/11	.0631
7	6/36	1	.1667
8	5/36	5/11	.0631
9	4/36	4/10	.0444
10	3/36	3/9	.0278
11	2/36	1	.0556
12	1/36	0	0
Total	1		.493

This means that you have a .493 probability of winning a dollar when you make a pass line bet and a .507 probability of losing a dollar so the expected value of a $1 bet is (.493)($1) + (.507)(-$1) = -$.014. Instead of losing 5 cents on average like you will at the roulette table, you only lose about a penny and a half per dollar pass line bet at the craps table.

It appears that craps will give us a better chance than roulette because of the better expected value. If you start making $1 pass line bets, how likely is it that your bank roll will get from $60 to $100 instead of running out of money? This question is one of many that are referred to as the Gambler's Ruin Problem.

If you are playing a fair game, i.e. .5 probability of winning $1 and .5 probability of losing $1 each round, your bank roll would be doing what is referred to as a symmetric random walk. I am excluding the proof but the likelihood of reaching your goal of $N when you start off with $i is equal to i/N. That is, if you were playing a fair game and you started with a bankroll of $60, there is a .6 probability that your bank roll would get to $100 before you ran out of money.

We are not, however, playing fair games. I'm again excluding the proof, but for asymmetric games, the probability of getting to your goal $N when you start with a $i when making $1 bets is equal to:

$$\frac{1 - (\frac{q}{p})^i}{1 - (\frac{q}{p})^N}$$

where p is the probability of winning and q is the probability of losing. Since $\frac{1-(\frac{.507}{.493})^{60}}{1-(\frac{.507}{.493})^{100}} \approx$ .2826, that means that there is about a 28% probability that you will get to $100 before running out of money by making pass line bets at craps. That might sound bad but it's much worse if you play roulette. Making $1 bets on red in roulette gives you only a 1.4% probability of getting the $100 necessary for the concert tickets before you go broke.

Is 28% the best we can do? I'm hoping that you realize that, when you don't have the edge, it is not a good strategy to have a lot of trials. While you are unlikely to come out ahead of Vegas in almost any situation, it is almost guaranteed that you will not if you have many trials because of the Law of Large Numbers. So what if you bet $40 instead of $1? You could bet the whole $60 but the goal is to get to $100 so you should not risk going broke on your first bet when winning $40 is all you need. This strategy is clearly better. On your first bet alone you have a 49.3% chance of reaching your goal! If you lose, all hope is not lost. You can bet all $20 again (remember we are trying to keep the number of trials as small as possible so we want to bet big). If you win, you're back to $40. Bet it all again. If you win, you're up to $80. Bet the $20 you need to get to $100. If you win, you get the concert tickets. If you lose, you're back where you started, with a $60 bankroll. Let's call the function a(x) the probability of reaching your goal when you have x dollars. We know we're a winner if we win on the first bet which has a 49.3% probability. We also get to $100 if we lose-win-win-win which has a probability of $(.507)(.493)^3 \approx .0608$. If we lose-win-win-lose we are back where we started and that has a probability of $(.507)^2(.493)^2$. Using this information we can set up the following equation to find the probability of getting to $100 when starting with $60.

$$a(60) = .493 + (.507)(.493)^3 + (.507)^2(.493)^2 a(60)$$

Solving for a(60) gives us a probability of approximately .5909 which means that using this strategy at craps would get the bankroll to $100 about 59% of the time!

Did we just figure out a way to beat the casino? Why or why not?

Losses are more significant than the gains

Homework Assignment 8.7: What about the Martingale? (in a row)

You start using the method discussed in Class Activity 8.8 to try to turn your $60 into $100. About 20 minutes into gambling, one of your classmates runs to you in the casino with an excited look on his face. He found out that someone was selling a $60 ticket online for $64. The person had originally bought the ticket hoping to sell it for a large mark-up, but since the concert is starting in a few hours he just wants to make a small profit on the transaction and will be happy to sell the $60 ticket for $64 and make four dollars.

One strategy that could be used to try to make $4 is the simplest of the Martingale betting strategies. To use this strategy the classmate will first bet $4. Let's say he bets on craps since we have seen that the probability of winning is greater in craps than it is in roulette. If he wins, which has a .493 probability, he can go to the show! If he loses, he doubles his bet to $8. If he wins on the second round he has still made a net profit of $4 and can go to the show. If he loses, he doubles his bet to $16. Again, if he wins, he can go to the show. If he loses, he doubles his bet one last time to $32. If he wins this time, he still has made $4 net profit and can use his $64 to buy the ticket. In fact, the only way that the classmate doesn't make the four dollars necessary is if he loses *four times in a row*. The probability of losing four times in a row is $(.507)^4 \approx .066$. This means that there is a 1 - .066 or 93.4% chance that the classmate wins four dollars and can go to the show.

Everything in the last two paragraphs is true. If that's the case, is this Martingale betting system a viable method for beating the house in Vegas? Explain your answer.

No, it is not a viable method for beating the house because of Gambler's Fallacy. It's critical to understand that each trial is independent.

It is only with unbounded wealth & time that the Martingale effect could be argued to be employed.

expected value = .978 - 5.74
= - EV = DON'T BET

170

Class Activity 8.10: Kelly Criteria: What to Do When You Have the Edge

So far we have only dealt with betting scenarios where your expected value is negative. These are normally all you will find in a place like Las Vegas because a negative expected value for you means a positive expected value for the casino and if they don't have a positive expected value they are not going to keep the lights on very long. Unless you are in one of the convoluted scenarios discussed in the last two units, the smartest thing you can do financially is to simply choose not to gamble.

Outside of Las Vegas, however, there are scenarios where you may have favorable odds and will need to decide how best to approach betting/investing. You may think that once you have an edge, even a small one, you're all set. Even if your odds are favorable you need to figure out what size bet or investment is optimal.

Let's imagine a betting or investing scenario where you have a 51% chance of winning $X and a 49% chance of losing $X. In addition, let's say that you have a bankroll of $1000. If you are going to make repeated bets or investments in this scenario where p = .51, how much should you bet? If you bet too little, your bankroll won't grow very fast, e.g. a $1 bet would only give you an expected value of 2 cents. If you bet too much, losses will severely cripple your bankroll, e.g. if you bet everything, five wins followed by one loss would leave you with nothing. What fraction of your bankroll do you think you should bet in order to maximize your long-run expected wealth at relatively low risk? _____

To maximize long-term growth, the Kelly criteria say that you should bet your edge, i.e. the proportion of your bankroll that you should bet is equal to p – q where p is the probability of winning and q is the probability of losing. If that's the case, what proportion of your bankroll would the Kelly criteria recommend you bet if you have a 51% chance of winning? _____

Homework Assignment 8.8: Chapter 8 Review

On October 7, 2015, the Powerball changed its format. Players previously chose five different numbers between 1 and 59 and a Powerball number between 1 and 35. Under the new system, players chose five different numbers between 1 and 69 and a Powerball number between 1 and 26.[39]

If a person bought a ticket under the old system, what was the probability it matched all six winning numbers?

$$_{59}C_5 = P_{1-59} = 5006386$$
$$_{35}C_1 = P_{PB} = 35$$
 } independent

$$5006386 \cdot 35 = 175223510$$

$$\boxed{\dfrac{1}{175223510}}$$

If a person bought a ticket under the new system, what was the probability it matched all six winning numbers?

$$_{69}C_5 = P_{1-69} = 11238513$$
$$_{26}C_1 = P_{PB} = 26$$
 } independent

$$11238513 \cdot 26 = 292201338$$

$$\boxed{\dfrac{1}{292201338}}$$

The Red Sox and Yankees play a best-of-seven series for the ALCS and in each game, the Yankees have a .55 probability of winning. What is the probability that the series goes to seven games? Dependent after game 4 *

$$\underbrace{(_6C_3)(0.55^7)}_{Yankees} + \underbrace{(_6C_3)(0.45^7)}_{Sox} = 0.379$$

(probability) * (probability)

$$\boxed{37.922\%}$$

You are playing a game of Texas Hold 'Em and the turn (the 4th community card) has just been revealed. You and one Amarillo Slim are the only two players left in the hand. Your pocket cards are an ace of clubs and a 4 of clubs. The flop (the first three community cards) was a 7 of clubs, a 9 of spades and a 10 of clubs. The turn was a 9 of diamonds. After the turn, Amarillo went all in. Based on what you know about Amarillo, you think that he has three 9s. The pot was at $140 and then Amarillo put in $40 to go all in. Let's assume that you are correct about Amarillo's hand and that the only way you will win the hand is if the river (the 5th and final community card) is a club that is not a 9 to give you a flush and not give Amarillo 4 of a kind. What decision has the greater expected value, folding or calling?

__folding__

An opinion poll asked an SRS of 1200 adults in Davidson County, "Do you think the Titans will win a Super Bowl by 2030?" Let's suppose that the population proportion of Davidson County who thinks "yes" is p = .3. If this opinion poll were repeated many times, the statistic $\hat{p}$ would be approximately normally distributed with mean .3 and standard deviation .013. Sketch the normal curve of this sampling distribution below.

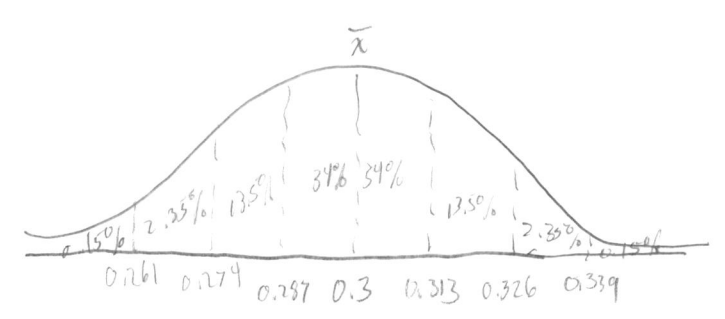

Using the 68-95-99.7 Rule, find the probability that between 30% and 32.6% of those polled say they think the Titans will win a Super Bowl by 2030.

34 + 13 + 5 = 47.5%

__0.475__

Part D: Inference

Chapter 9. Confidence Intervals for Proportions

Class Activity 9.1: A More Perfect Confidence Interval

In chapter 2, we learned that a quick method for calculating margin of error with 95% confidence is to find $\frac{1}{\sqrt{n}}$ where n is the sample size under the assumption that the population size is at least ten times greater than the sample. Now that we have become more comfortable working with distributions and have a more complete understanding of normal distributions and sampling distributions, we can calculate margins of error more accurately and with any degree of confidence that we would like.

In chapter 8, when we discussed sampling distributions, the mean and standard deviation of the distribution of $\hat{p}$ were given. In general, if a large SRS of size *n* is taken from a large population where the truth about the population is some parameter *p*, the sampling distribution of $\hat{p}$ is 1, approximately normal 2, has a mean of p and 3, has a standard deviation of $\sqrt{\frac{p(1-p)}{n}}$.

Let's say that 70% of people who use the internet use Chrome as their browser. If we take an SRS of 3500 internet users, ask what browser they use and divide by 3500 we will get some sample statistic $\hat{p}$. If we took a sample this way many times, we could produce a sampling distribution of $\hat{p}$. The sampling distribution would be close to normal. What would the mean and standard deviation of the distribution be?

_____

Using the 69-95-99.7 Rule, between what two proportions would you expect the middle 99.7% of $\hat{p}$ values to fall if many samples were taken?

_____

Homework Assignment 9.1: Where Did $1/\sqrt{n}$ Come From?

The 68-95-99.7 Rule says that in a normal distribution, the middle 95% of observations fall within two standard deviations of the mean. *That means that the mean is within ± two standard deviations of the middle 95% of observations.* Let's use our new formula to find the margin of error necessary to capture the parameter value p 95% of the time *if p is .5*. The margin of error is going to be double the standard deviation because you need two standard deviations to capture the middle 95% of observations in a normal distribution. This means that margin of error is going to equal $2\sqrt{\dfrac{p(1-p)}{n}}$. Set p equal to .5 and simplify the expression as much as possible below.

$$\text{margin of error} = 2\sqrt{\dfrac{p(1-p)}{n}} = \dfrac{2(0.5)}{\sqrt{n}}$$

$$= \boxed{\dfrac{1}{\sqrt{n}}}$$

Class Activity 9.2: What Happens When p ≠ .5?

Using our new formula for standard deviation of p, $\sqrt{\dfrac{p(1-p)}{n}}$, let's calculate the standard deviation of p in two different distributions.

Situation 1, an SRS of 100 is taken from a population where p = .5:

Situation 2, an SRS of 100 is taken from a population where p = .9:

What happens to the standard deviation of p when p or 1-p is close to 0?

_____

To understand why this is the case, let's imagine that in situation 1, we are taking a SRS of 100 people from the population of Americans and recording the gender of the individuals selected. For the purpose of this example, let's say that the population parameter p of those who are male is .5. If the result of our sample was that $\hat{p}$ = .6 or .4, i.e. we ended up with 60 males and 40 females in our sample or vice versa, would that be surprising?

How many standard deviations away from our expected proportion would that sample statistic be?

_____

What percent more males were there in our sample than we expected if $\hat{p}$ = .6 and what percent fewer males were there in our sample than we expected if $\hat{p}$ = .4 (it's the same answer)? _____

Now, let's imagine that in situation 2, we are taking a SRS of 100 people from the population of Americans and recording if the person is left-handed. For the purpose of this example, let's say that the population parameter p of those who are left-handed is .1. If the result of our sample was that $\hat{p}$ = 0 or .2, i.e. we ended up with 0 left-handed and 100 right-handed people in our sample, or 20 left-handed people and 80 right-handed people, would that be surprising?

How many standard deviations away from our expected proportion would that sample statistic be?

_____

What percent more left-handed people were there in our sample than we expected if $\hat{p}$ = .2 and what percent fewer left-handed people were there in our sample than we expected if $\hat{p}$ = .0 (it's the same answer)? _____

Why does standard deviation of the sampling distribution of $\hat{p}$ decrease when p or 1 – p approaches zero? _____

Class Activity 9.3: What If We Don't Know p?

While using our new formula for the standard deviation of p, you might have asked the following question: If I know the value of p, why would I be taking a sample? It's a very legitimate question. In most cases, a sample is taken to get a statistic value $\hat{p}$ that gives us some idea what the population parameter p might be. For example, if I know that 10% of the population of Americans is left-handed (p), why would I want to take a sample to get a statistic value that helps me create an interval where I believe p may be found?

As a result, we need some way to find an approximate standard deviation when taking a sample when we don't know the value of p. What value could replace p? How about $\hat{p}$? While it is not perfect, using $\hat{p}$ as an approximation for p allows us to create a confidence interval even when p is unknown. This means that an approximate 95% confidence interval can be found with the following formula:

$$\hat{p} \pm 2\sqrt{\frac{\hat{p}(1-\hat{p})}{n}}$$

$\hat{p}$ is used in place of p to get an approximate standard deviation and the standard deviation is multiplied by two because the 68-95-99.7 rule tells us that the middle 95% of observations in a normal distribution are within two standard deviations of the mean.

Note: some textbooks refer to the estimated standard deviation $\sqrt{\frac{\hat{p}(1-\hat{p})}{n}}$ as the standard error.

Homework Assignment 9.2: A More Accurate 95% Confidence Interval

Mr. Shone takes a random sample of MBA graduates and performs a survey. He asks 645 graduates: "If it were your decision, would Latin be a required course at MBA?" 517 of the 645 graduates polled said "yes" they would have Latin be a required course at MBA.

Explain in words what p means in this setting.

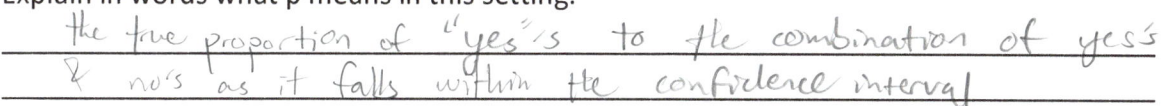

the true proportion of "yes"'s to the combination of yes's & no's as it falls within the confidence interval

Use this survey result to find a 95% confidence interval for p.

517/645 = $\hat{p}$ = 0.8016

0.8016 + 1.96 √(0.8016(1-0.8016)/645) = 0.8324

to

0.8016 − 1.96 √(0.8016(1-0.8016)/645) = 0.7708

Class Activity 9.4: Different Levels of Confidence

While 95% confidence has become the standard level of confidence, there may be situations where you would prefer to have more or less confidence in your interval. In our new 95% confidence interval formula, we multiplied the standard deviation by 2 because the middle 95% of observations are contained within approximately 2 standard deviations. If we want a different level of confidence, we simply need to find the number of standard deviations that we must add to and subtract from the mean to capture our desired proportion of observations. We call this desired number of standard deviations, the **critical values** of the normal distributions and we use the symbol z^* to represent critical values.

This means that our formula for an approximate level C confidence interval for p is found with the formula:

$$\hat{p} \pm z^* \sqrt{\frac{\hat{p}(1-\hat{p})}{n}}$$

This formula can be used when an SRS of some large value n is taken from a population. Some popular confidence levels and their corresponding critical values z^* can be found in Table 9.1 below. Note: while we have been using 2 for 95% confidence, this table shows that 1.96 is a more accurate critical value z^*. For reference, this table can also be found as Table B at the end of the book.

Confidence Level C	Critical Value z^*	Confidence Level C	Critical Value z^*
50%	0.67	90%	1.64
60%	0.84	95%	1.96
70%	1.04	99%	2.58
80%	1.28	99.9%	3.29

Table 9.1

If Mr. Shone wanted a 99% confidence level when reporting the results of his Homework Assignment 9.2 Latin survey, what would his confidence interval be?

Class Activity 9.5: Finding Levels of Confidence

The eight critical values given in the table are a good start, but what if you want a 99.999%, 75%, or 20% confidence level? You could look for a bigger table but it would be better to learn how to derive z* so you could find a z* for any confidence interval. To see how to do this, let's say we want to find the z* that corresponds to a 65% confidence level. This really means that we want to find the number of standard deviations that need to be added to and subtracted from the mean to capture the middle 65% of the observations in a normal distribution. In Figure 9.2 below, the middle 65% of observations are the light blue region in the center of the bell curve.

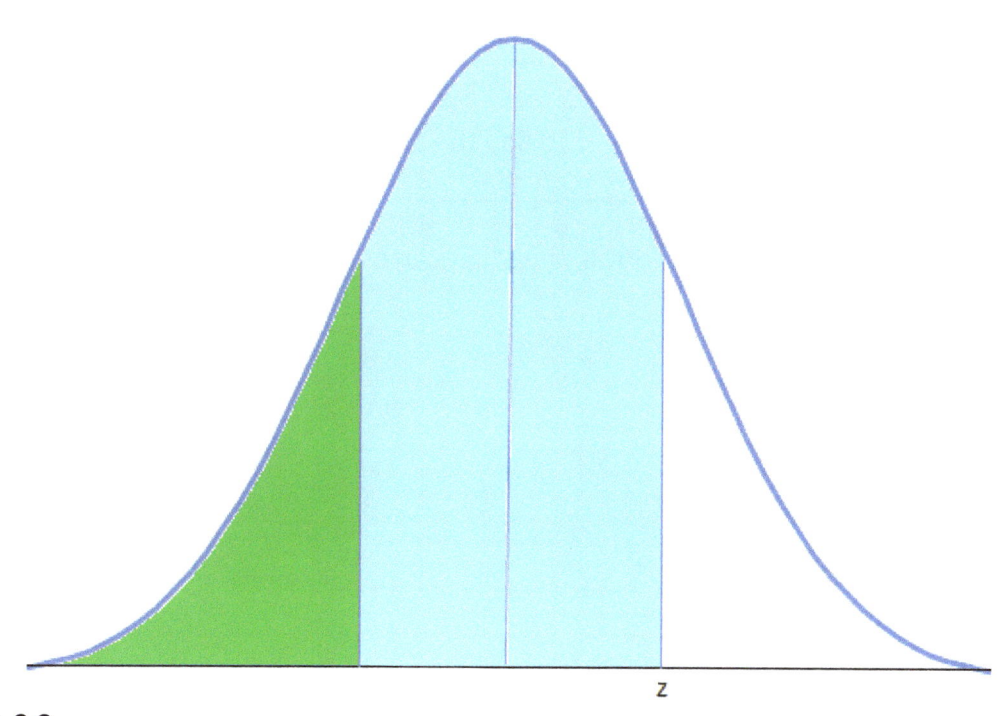

Figure 9.2

We can find standard normal probabilities, i.e. the proportion of values to the left of an observation with a given z-score in a normal distribution, using Table A or our calculator. *The z* value that would capture the middle 65% of observations would be equal to the z-score that has .825 values to the left in our standard normal probability table.* The change from 65% to .825 is the result of adding the green portion because standard normal probabilities capture *all* observations to the left of a given z-score. If 65% of observations are within z* standard deviations, 35% are outside z* standard deviations and half of that,

181

17.5%, are to the left outside z* standard deviations. 65% + 17.5% = 82.5% = .825. Using our Table A, we can see that z* for a confidence level of 65% should be between .9 and 1. If we want a more accurate z*, we can use the invNorm command (2nd, VARS, 3) on our calculator. If we do, we get a z* of ≈.935 as seen below.

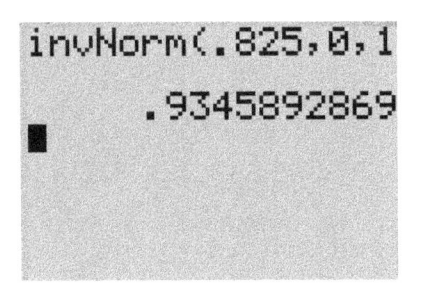

```
invNorm(.825,0,1
            .9345892869
```

What would you input on your calculator to find the z* critical value that corresponds to a confidence level of 99.99%? _____

Use your calculator to find the z* critical value that corresponds to a confidence level of 99.99%. _____

Homework Assignment 9.3: Captain Norton and Service

Captain Norton wonders how the student body feels about compulsory service. He selects an SRS of 70 current MBA students and asks the question: "Do you support service being compulsory at MBA?" Of the 70 students asked, 50 said that yes, service should be compulsory. Captain Norton wants to present his findings at MBA's next board meeting. He would like to make a confidence statement with a confidence level of 98%. Find the z* necessary for 98% confidence and then find the confidence interval for Captain Norton's study.

$$\hat{p} = \frac{50}{70} = 0.714$$

$$\boxed{z^* = 2.333}$$

$$0.714 + 2.333\sqrt{\frac{0.714(1-.714)}{70}}$$

$$0.714 - 2.333\sqrt{\frac{0.714(1-.714)}{70}}$$

$$\boxed{0.1694 \text{ to } 0.872}$$

Tricky one: Captain Norton takes an SRS and asks the follow up question: "Do you support at least 50 hours of service a year being compulsory at MBA?" 20% of students polled said yes. How many students need to have been in that sample for Captain Norton to be *at least* 90% confident that between 15% and 25% of the population support 50 hours of compulsory service a year at MBA?

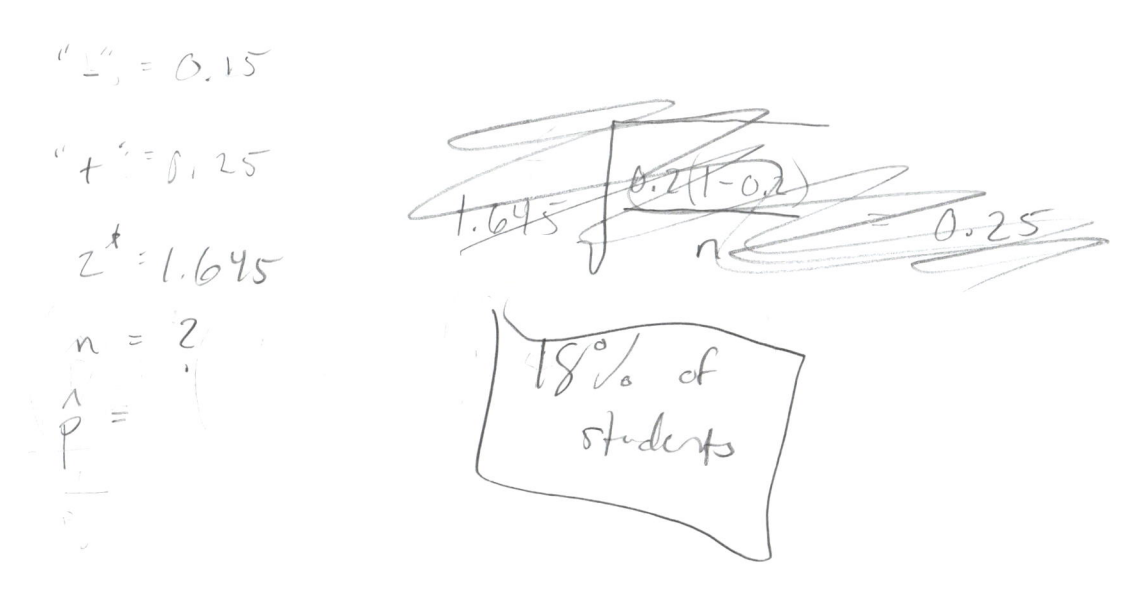

"−" = 0.15

"+" = 0.25

$z^* = 1.645$

$n = ?$

$\hat{p} = $

$$1.645\sqrt{\frac{0.2(1-0.2)}{n}} = 0.25$$

$$\boxed{18\% \text{ of students}}$$

Chapter 10. Significance Tests and Confidence Intervals for Means

Class Activity 10.1: Free Throws at the Faculty-Student Game

The week before the faculty-student basketball game, Mr. Davidson makes a point of bragging about his abilities on the court. "While most centers struggle from the charity stripe," Mr. Davidson explains, "I am a respectable 70% free throw shooter." The students are fed up with Mr. Davidson's talk so they decide to put his free throw claim to the test. At the faculty-senior game, the students employ a Hack-A-Davidson strategy that sends Mr. Davidson to the free throw line 15 times. Mr. Davidson makes 6 of his 15 free throw attempts in the game.

The next day in class, the students claim that they proved that Mr. Davidson was lying and that he is not, in fact, a 70% free throw shooter. "Far from it," says Mr. Davidson. "My shooting 6 for 15 from the free throw line does not prove I am a worse than 70% free throw shooter any more than a fair coin's string of 6 heads in a row does not prove that the coin is unfair. Some days I shoot 100% and some days I shoot 40%. All you witnessed was one of the less impressive shooting performances put on by a true 70% free throw shooter." The students are momentarily stymied. Is what Mr. Davidson saying true? In one sense, it is true. Is it possible that Mr. Davidson is a 70% free throw shooter and it just happened to be one of his worst days at the line? Yes. Is it probable? No. In the same way, if you flip a coin 10 times in a row and it comes up heads every time, is it possible that it's a fair coin? Sure. Is it probable? No.

How could the students argue that Mr. Davidson's claim is false? One of the most popular forms of argument is the *reductio ad absurdum (reductio* for short), which translates to "reduction to absurdity." A reductio argument starts off by assuming some premise is true, proves that that premise leads to a contradiction or an absurd conclusion, and in doing so proves that the assumed premise must in fact be false. In statistics, we can perform a **significance test**, a process that determines that probability that an observed sample result occurred by chance given some assumption. We call the assumption the **null hypothesis (H_0)**. We call the probability that a sample effect as or more extreme than the one observed occurred just by chance the **P-value**. The smaller the P-value, the stronger our evidence is to reject the null hypothesis in favor of the **alternative hypothesis (H_a)**. The alternative hypothesis represents the opposite of the null hypothesis. The alternative hypothesis is true if the null hypothesis is false. Unlike a reductio, we do not prove that the null hypothesis is true or false with a significance test, instead we can show that the null hypothesis is very unlikely given the evidence available. As Jordan Ellenberg referred to significance testing, it is "[n]ot a reductio ad absurdum...but a reductio ad unlikely."[40]

How would this kind of significance test look in practice? In the case of the coin landing heads 10 times in a row, the students could make the following argument: 1) *Let's assume the coin is fair,* 2) *If the coin is fair and I flipped it 10 times, it would land heads 10 times in a row (1/2)¹⁰ = 1/1,024 of the time.* 3) *Since a result that is so different from my expected value of 5 heads and 5 tails would occur so infrequently, it might be more reasonable to assume that the coin is not fair.* What about Mr. Davidson's free throws? The students could do the same thing. The students could assume that Mr. Davidson is in fact a 70% free throw shooter and, *given that assumption*, find out how often Mr. Davidson would make 6 or fewer free throws when he shoots 15. If it is highly unlikely, there is strong evidence against the original assumption that Mr. Davidson is a 70% free throw shooter.

To find the P-value, the probability that Mr. Davidson would make 6 or fewer free throws when he shoots 15 if he really is a 70% free throw shooter, we are going to use a **Z-test**. A Z-test is a significance test that finds the likelihood of a value as extreme or more extreme than a given statistic value occurring within a normal distribution under the assumption of a null hypothesis. Back in chapter 9, I said that we don't use formula $\sqrt{\frac{p(1-p)}{n}}$ for standard deviation very often because if p were known, we normally would not bother talking about samples and statistic values within a sampling distribution. In this case, however, this formula comes in handy because we are operating under the assumption that we do know p. That is, we are assuming that our null hypothesis that Mr. Davidson is telling the truth and he does in fact shoot 70% from the free throw line is true.

If the null hypothesis is true, the mean of the sampling distribution of $\hat{p}$ is .7 and the standard deviation is $\sqrt{\frac{.7(1-.7)}{15}} \approx .12$.

What is the value of $\hat{p}$ that was observed at the Faculty-Student basketball game? _____

Sketch the sampling distribution of $\hat{p}$ below and label the observed $\hat{p}$ from the Faculty-Student game.

How many standard deviations below the mean is $\hat{p}$? _____

Use Table A at the end of the book to determine the probability of Mr. Davidson shooting that poorly or worse when shooting 15 free throws if he is in fact a 70% free throw shooter. This is the P-value. _____

To say that a sample result is *statistically significant at the x% level* means that it could occur by chance no more than x% of the time in repeated samples. It is true when the P-value is smaller than x. We call this x% level the **significance level**, and it is normally written as α, the Greek letter alpha. α = .05 is an arbitrary but commonly used significance level.

Was Mr. Davidson's 6 for 15 free throw shooting performance significant at the α = .05 level? ____ The α = .01 level? ____ The α = .005 level? ____

Class Activity 10.2: Tap Water vs. Lime Spa Water: One-Sided and Two-Sided Alternative Hypotheses

Mr. Davidson is intrigued by the spa water in the Dining Hall and he wonders if people like unflavored tap water more or less than the lime spa water. In the free throw shooting example, the null hypothesis H_0 was that p = .7 (that Mr. Davidson makes 70% of his free throws), and the alternative hypothesis H_a was that p < .7 (that Mr. Davidson makes fewer than 70% of his free throws). This H_a is an example of a **one-sided alternative hypothesis**. A one-sided alternative hypothesis calculates a P-value by looking at the probability of getting a result as extreme as the sample result in one direction.

Our null hypothesis in the water study would be that lime spa water and tap water are liked equally, or that p = .5 where p is the proportion that prefer lime spa water. However, unlike the free throw example in which we really thought that Mr. Davidson was overstating his ability, it does not make sense to say that our alternative hypothesis is p > .5 or p < .5. We're really not sure if there is a difference *in either direction*. This means that this situation would be a good candidate for a **two-sided alternative hypothesis**. A two-sided alternative hypothesis calculates a P-value by looking at the probability of getting a result as extreme as the sample result in both directions. The two-sided alternative hypothesis here would be written as p ≠ .5.

Let's say that Mr. Davidson took a random sample of 100 individuals and had them taste lime spa water and tap water and say which water they preferred. 65 of the 100 in the sample said that they preferred the lime spa water. The sampling distribution of $\hat{p}$ given that our null hypothesis is true would have a mean of .5 and a standard deviation of $\sqrt{\frac{.5(1-.5)}{100}}$ = .05. Our $\hat{p}$ is .65 (65/100).

How many standard deviations above the mean is our $\hat{p}$ (what is its z-score)?

_____

What is the P-value, i.e. what is the probability of getting a same result this many standard deviations above the mean *or this many standard deviations below the mean* if the null hypothesis is true?

_____

Homework Assignment 10.1: Finding P-values on Your Calculator

The calculator is a great tool when performing significance tests, especially as the tests get more and more complicated. They can save a lot of time and can give more accurate P-values that those found using tables. Let's find the P-value in our tap water vs. lime spa water study. On your calculator, Press STAT, choose TESTS, and then select 5: 1-PropZTest. Enter .5 for P_0 to show that the null hypothesis is p = .5, enter 65 for x to represent the number of people in the sample who said they prefer lime spa water, and enter 100 in for n to represent the sample size. By default ≠P_0 should be selected which indicates that we have a two-sided alternative hypothesis. If you've entered this information in correctly, your screen should look like the following.

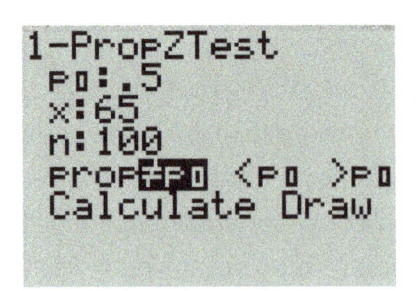

If you select the "Calculate" option, the calculator will find the statistic value $\hat{p}$, the z-score of the statistic, and the P-value. If you select the "Draw" option, the calculator will give you the z-score of the statistic, the P-value, and will draw a normal distribution and fill in the area corresponding the probability of getting a result as extreme as or more extreme than the sample result. What was the P-value found by the calculator in the water study?

.0027

Let's do the same thing for our Mr. Davidson shooting free throws example. This time your screen should look like the following.

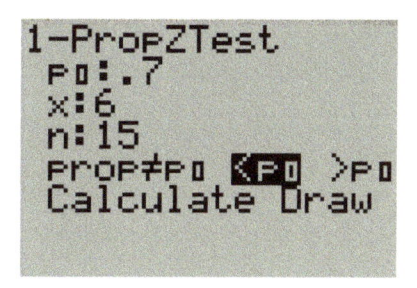

```
1-PropZTest
  p0:.7
  x:6
  n:15
  prop≠p0 <p0 >p0
  Calculate Draw
```

What P-value does the calculator find in the free throw example? __0.9987__

Mr. Davidson makes one last challenge to the students. "Shooting free throws is a binomial setting," he begins. "Normal distributions should only be used for binomial approximations when there is a large sample size. I only shot 15 free throws and as a result we cannot trust your P-value." Mr. Davidson is correct, which the 100 sampled in the water study were certainly enough to use a normal approximation, a sample size of 15 is very small. In Figure 10.1 below, you can see how the normal distribution we used compares to the actual binomial distribution you would get with a probability of success equal to 70% and a sample size of 15.[41]

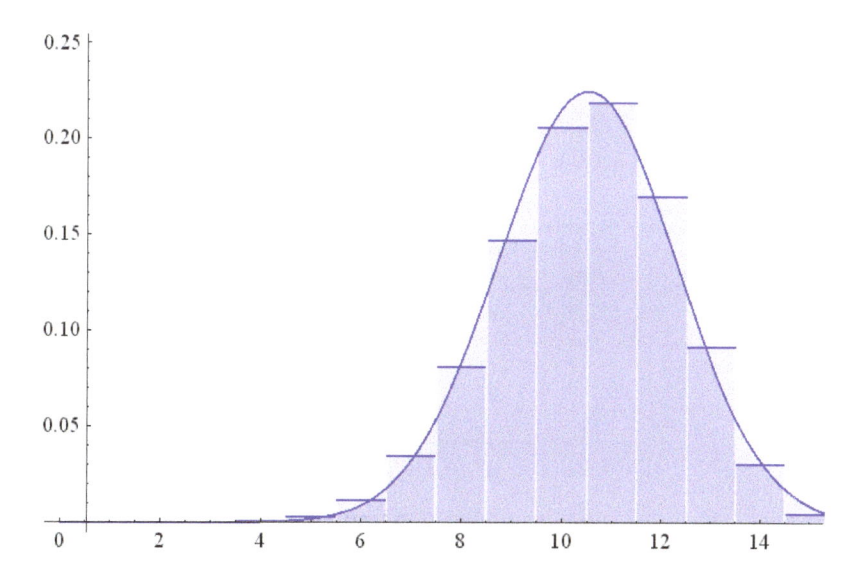

Figure 10.1 Normal vs. Binomial Distribution with p = .7 and n = 15

What could the students do to get a more accurate P-value. One option is to do a continuity correction. As you can see in Figure 10.2, it would be a better estimate of the binomial probabilities to find the area to the left of 6.5 instead of the area to the left of 6. Therefore, we could use a $\hat{p}$ of .43 (6.5/15) instead of .4 (6/15) to find our z-score and corresponding P-value.

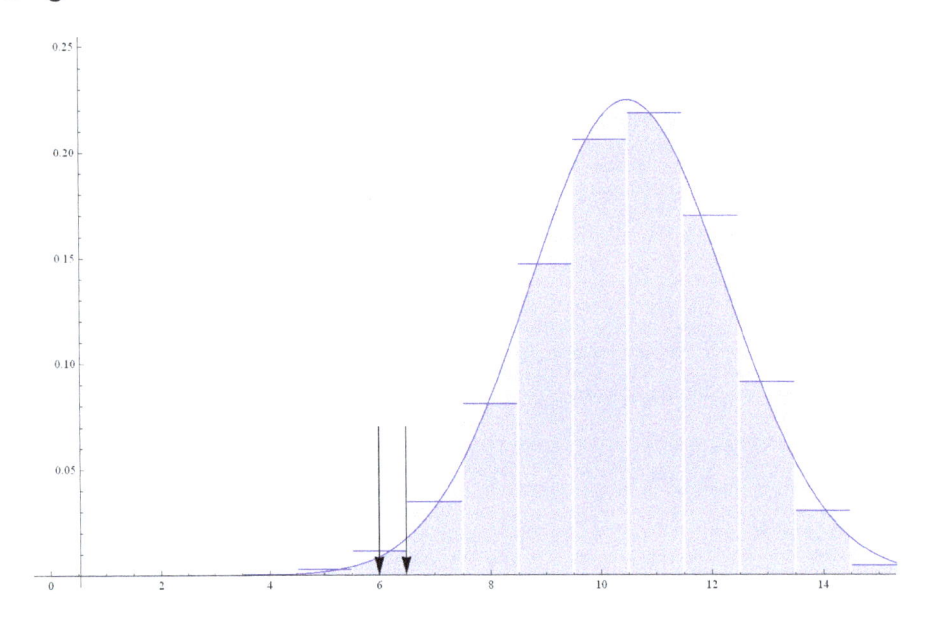

Figure 10.2 Continuity Correction for Binomial to Normal Distribution

Or, we can just find the exact probability of making 6 or fewer free throws by using binomcdf. Press 2nd, VARS and then select A:binomcdf(from the list. Type "15" for the number of trials, a comma, ".7" for the probability of making a free throw, another comma, and finally "6" in order to find the probability of making 6 or fewer free throws. When you are finished your calculator should look like this.

What is our new and more accurate P-value? .0152

Was Mr. Davidson's 6 for 15 free throw shooting performance significant at the α = .05 level? Yes The α = .01 level? No

Class Activity 10.3: Testing Our Own Free Throw Claims

We are now going to perform a Z-test to test the hypotheses relating to the free throw percentage of students in the class. Each student in the class is going to shoot 50 free throws. Before we shoot the free throws you need to make some claim about your free throw percentage. To make the math easy, you are going to have 4 options for your null hypothesis: p = .2, p = .4, p = .6 or p = .8 where p is your shooting percentage. Yet again we are making the assumption that free throws are independent random events. Since we do not know if you might be understating or overstating your free throw shooting percentage, we are going to use a two-sided alternative hypothesis. State your null and alternative hypotheses below.

H_0: $0.6 = p$

H_a: $p \neq 0.6$

How many free throws did you make out of 50? __27__

Mean = 0.6

Homework Assignment 10.2: What Can We Conclude About Our Free Throw Shooting?

Sketch the sampling distribution of $\hat{p}$ below assuming H_0 is true and label the observed $\hat{p}$ from Class Activity 10.3.

$\hat{p} = \frac{24}{50} = 0.48$

$\sqrt{\frac{\hat{p}(1-\hat{p})}{n}}$

0.6

0.0693

p = 0.0587

How many standard deviations above or below the mean was $\hat{p}$, i.e. what was $\hat{p}$'s z-score?

$\frac{x - \mu}{SD}$ $\frac{\hat{p} - p}{SD} = \frac{0.54 - 0.6}{0.0693}$

$\frac{27}{50}$

__−0.8658__

Find the P-value using our standard normal probabilities table.

$z = \frac{0.048 - 0.6}{0.0693}$

29.5

0.386

Find the P-value using 1-PropZTest on your calculator.

0.386

Is there enough evidence to reject the null hypothesis H_0 at the α = .05 level? __No__

When our P-value is less than our α value we say that we reject the null hypothesis at that α level. This means that the effect observed is so great that it would occur by chance less than α and therefore we believe H_0 is false. When our P-value is more than our α value, it is a common mistake to say that we have *proved the null hypothesis. This is incorrect.* Instead, if the P-value is greater than our α value, we say that we *failed to reject the null hypothesis.* If your null hypothesis was that you are a 60% free throw shooter and you made 31/50 or 62% of your free throws you should definitely not reject your null hypothesis, but you have not proven that you are a 60% free throw shooter.

To understand why, consider the parallel of a courtroom. We say that a person is "innocent until proven guilty" and then say that the prosecution must prove it "beyond a reasonable doubt." That means that, as a hypothesis test, our null hypothesis is that the person is innocent and the alternative hypothesis is that the person is guilty. If the prosecution is successful in convincing the judge or jury that, given the evidence provided, there is a very small chance (say less than 1%) that the person is innocent then the person's innocence may be rejected in favor of the alternative hypothesis that the person is guilty. If the prosecution is unsuccessful, however, that does not prove that the person is innocent. All we know is that there was not enough evidence found to confidently state that he was guilty. There are many guilty men who were not convicted of crimes they committed. There are many more null hypotheses that were not rejected and nonetheless are not true.

When a null hypothesis is false but we fail to reject it, we say that a **Type II error** has occurred. When a null hypothesis is true but reject it, we say that a **Type I error** has occurred. The probability of a Type I error is α. The probability of a Type II error is known as β, and the **power of the hypothesis test** is $1 - β$.

Let's discuss these new terms in the context of Mr. Davidson's claim that he is a 70% free throw shooter. If we have Mr. Davidson shoot free throws and then perform a Z-test with the null hypothesis that Mr. Davidson's claim is true, there is a risk of a Type I error and a risk of a Type II error. If we use an α of .05 that means there is a 5% chance that we reject Mr. Davidson's null hypothesis that he is a 70% free throw shooter even though he is a 70% free throw shooter. That is an example of a Type I error. That means that, since we are using an α of .05, there is a 5% chance that the class calls Mr. Davidson a liar when in fact he is a 70% free throw shooter. If the class is worried about wrongly accusing Mr. Davidson of being a liar, they could lower the α level to .01 or, if you want to be really confident that Mr. Davidson is lying, .001.

A small α sounds great, right? It does decrease the risk of a Type I error but what's the trade-off? It increases the risk of a Type II error. We won't cover calculating β in this text, but it does increase as α decreases. That means that if α is .001, there is a good chance that Mr. Davidson's null hypothesis is not true but he won't get caught. What's the class to do? The students don't want to wrongly accuse Mr. Davidson of lying but they also want to catch him if he is misrepresenting his ability. There is one solution. β values increase as α decreases, but β also responds to sample size, n. By choosing a small enough α the class can be confident beyond a reasonable doubt that it is not wrongly accusing Mr. Davidson of being a liar and by using a large enough sample size it can decrease the probability of a Type II error, β, which means that if Mr. Davidson is lying they are more likely to catch him. Why? Because 1 – β, the power of the hypothesis test, is the probability that a null hypothesis is rejected when it is false.

When in the context of challenging Mr. Davidson's basketball trash talk, finding appropriate α and β seems a lot of effort for something unimportant. Consider then, the use of significance tests on some new drug in a clinical trial. A Type I error in a clinical trial means that a new drug which is not actually effective is put on the market. A Type II error in a clinical trial would mean that an effective new drug is determined ineffective and is not made available to patients. It these types of situations, a deep understanding of Type I and Type II errors is critical.

On the comic on the next page, what type of error occurred which led to the conclusion that green jelly beans are causally connected to acne?[42] _____

JELLY BEANS CAUSE ACNE!
SCIENTISTS! INVESTIGATE!
BUT WE'RE PLAYING MINECRAFT!
...FINE.
WE FOUND NO LINK BETWEEN JELLY BEANS AND ACNE (P > 0.05).
THAT SETTLES THAT.
I HEAR IT'S ONLY A CERTAIN COLOR THAT CAUSES IT.
SCIENTISTS!
BUT MIIIINECRAFT!
WE FOUND NO LINK BETWEEN PURPLE JELLY BEANS AND ACNE (P > 0.05).
WE FOUND NO LINK BETWEEN BROWN JELLY BEANS AND ACNE (P > 0.05).
WE FOUND NO LINK BETWEEN PINK JELLY BEANS AND ACNE (P > 0.05).
WE FOUND NO LINK BETWEEN BLUE JELLY BEANS AND ACNE (P > 0.05).
WE FOUND NO LINK BETWEEN TEAL JELLY BEANS AND ACNE (P > 0.05).
WE FOUND NO LINK BETWEEN SALMON JELLY BEANS AND ACNE (P > 0.05).
WE FOUND NO LINK BETWEEN RED JELLY BEANS AND ACNE (P > 0.05).
WE FOUND NO LINK BETWEEN TURQUOISE JELLY BEANS AND ACNE (P > 0.05).
WE FOUND NO LINK BETWEEN MAGENTA JELLY BEANS AND ACNE (P > 0.05).
WE FOUND NO LINK BETWEEN YELLOW JELLY BEANS AND ACNE (P > 0.05).
WE FOUND NO LINK BETWEEN GREY JELLY BEANS AND ACNE (P > 0.05).
WE FOUND NO LINK BETWEEN TAN JELLY BEANS AND ACNE (P > 0.05).
WE FOUND NO LINK BETWEEN CYAN JELLY BEANS AND ACNE (P > 0.05).
WE FOUND A LINK BETWEEN GREEN JELLY BEANS AND ACNE (P < 0.05).
WHOA!
WE FOUND NO LINK BETWEEN MAUVE JELLY BEANS AND ACNE (P > 0.05).
WE FOUND NO LINK BETWEEN BEIGE JELLY BEANS AND ACNE (P > 0.05).
WE FOUND NO LINK BETWEEN LILAC JELLY BEANS AND ACNE (P > 0.05).
WE FOUND NO LINK BETWEEN BLACK JELLY BEANS AND ACNE (P > 0.05).
WE FOUND NO LINK BETWEEN PEACH JELLY BEANS AND ACNE (P > 0.05).
WE FOUND NO LINK BETWEEN ORANGE JELLY BEANS AND ACNE (P > 0.05).
NEWS
GREEN JELLY BEANS LINKED TO ACNE!
95% CONFIDENCE
ONLY 5% CHANCE OF COINCIDENCE!
SCIENTISTS...

Class Activity 10.4: Is Mr. Davidson's Die Fair?

So far our significance tests have applied to **binary variables**, variables that take one of two values, e.g. made free throw or missed free throw, preferred tap water or lime spa water. What if variables can take more than one value? Let's say for example that a student believes that Mr. Davidson's die might be unfair, i.e. that certain numbers occur with more regularity than others. To test this theory, he rolls the die 60 times and gets the following results.

Result	Frequency
1	9
2	6
3	11
4	19
5	9
6	6

The expected frequency or count for each number is 10, but we would expect to see some variation due to the randomness of rolling the die even if the die is fair. Do these results suggest that the die is unfair? We could use a Z-test for each number, e.g. H_0: p = 1/6, H_a: p ≠ 1/6 for 1, 2,…, and 6, but that would be inefficient and would not tell us how likely results like these six frequencies would occur if the die were fair. We need a new test, the **chi-square (χ^2) test for goodness of fit**. It is a significance test used to test one categorical variable from a single population to see if sample data is consistent with the distribution in the null hypothesis.

The chi-square (χ^2) test for goodness of fit, like all of our chi-square tests, uses the χ^2 test statistic which is defined as follows:

$$\chi^2 = \sum \frac{(observed - expected)^2}{expected}$$

In this problem, what are the smallest and largest values that χ^2 can take? _____ and _____.

When will χ^2 be large? _____

Let's find χ^2 for our observations.

χ^2 = _____

Is this value large or small? It's tough to say without more information. To get some idea, for your homework you are going to repeatedly simulate rolling a fair die 60 times. After 100 simulations and after calculating χ^2 after each trial, you should have some idea if our χ^2 value is large enough to think that the die is unfair, or small enough to attribute it to mere chance variation.

This simulation requires some more complex operations on our calculators, so we are going to do the first couple simulations in class to make sure everyone is successful on their calculators. For homework, each student will run the next 98 simulations and record the results below to make a histogram.

We are going to use L1 and L2 in our simulation, so our first step is clearing L1 and L2. Secondly, we are going to have our simulation keep track of which number simulation we are running so we do not lose track of where we are. The variable C is going to help us count the number simulation we just ran. To do this, we want C to currently equal 0. To make C equal 0, we use the store (STO→) button on our calculator. After you store 0 as C, your calculator should look like the following.

```
0→C
              0
```

The third step is where it gets complicated. We want the calculator to simulate rolling a fair die 60 times and to store those 60 values in L1. Then we want the calculator to add up how many 1s, 2s, 3s, etc. we got in our simulation and store the number of 1s, 2s, 3s, etc. in L2. Lastly, we want the calculator to find the χ^2 statistic value for that simulation and tell us what it is while also telling us what number simulation we just ran. If you want a calculator to perform multiple steps with the click of a button you can write a program, or you can separate the steps with colons. In this simulation, we will be using the latter method. The following shows what you should type into your calculator to run the simulation and shows the results of my first two simulations. Obviously since these are randomly produced results, you may perform the simulation correctly and get a different result for your χ^2 test statistic:

```
randInt(1,6,60)→
L₁:seq(sum(L₁=X)
,X,1,6)→L₂:sum((
L₂-10)²/10)→S:C+
1→C:{C,S}
                {1 11}
                {2 2.2}
```

For the homework tonight, you are going to hit ENTER 98 more times to run 98 more simulations. Every time you run a simulation, you should round the χ^2 test statistic to the nearest integer and plot it on the histogram below. For example, I would fill in the rectangle above 11 and above 2 after my first two simulations. You will use this histogram to help determine if we think our die is fair.

Homework Assignment 10.3: Making a Chi-Square Distribution for Five Degrees of Freedom

Fill in the rectangles corresponding to the 100 χ^2 statistic values you found when you ran your 100 die rolling simulations.

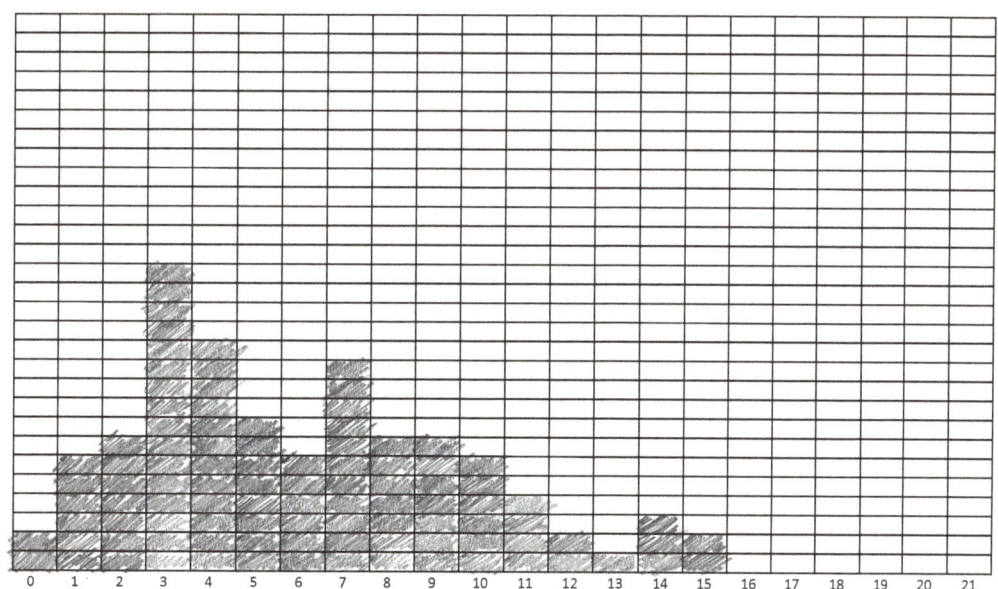

Is this distribution skewed right, skewed left, or approximately symmetric? _____
_____ skewed right _____

After rounding, how many χ^2 statistic values were larger than the χ^2 statistic we found for our data in Class Activity 10.4? ~~_____~~

Do you believe our die is fair or unfair? Why? Unfair; it's a program that makes outcomes based on algorythms; if you want a true set of outcomes without bias, roll a real die

Class Activity 10.5: Using the Chi-Square Distribution to Perform a Chi-Square Test for Goodness of Fit

The values that χ^2 will take depend on the number of categories our variable can take. When there are more possible categories, large χ^2 values occur more frequently. As a result, there is a family of χ^2 distributions and we identify different χ^2 distribution graphs according to the number of categories our variable can take. More exactly, we label them by our variable's **degrees of freedom**. Degrees of freedom = the number of categories – 1. The histogram we made for Homework Assignment 10.3 showed an approximation of the density curve for a χ^2 distribution with 5 degrees of freedom (6 possible numbers – 1).

Let's take a look at the exact density curve corresponding to the χ^2 distribution with df = 5. On your calculator, push Y= and in your Y1 graph type χ^2pdf(X,5). χ^2pdf(can be found by pushing 2nd, VARS. Your calculator should look like the following:

Before hitting graph, we want to choose an appropriate window to see our density curve. Push WINDOW and enter the following values.

Now if you press GRAPH you should see a density curve that looks similar to the histogram you produced in Homework Assignment 10.3.

I referred to the graph on the calculator as the exact density curve for the χ^2 distribution with df = 5. In reality, it is the density curve that is approached as sample size increases. Just like Z-tests, the χ^2 test for goodness of fit is only effective when sample size and thus expected counts are sufficiently large. A good rule of thumb is that all expected counts should be at least five if we want to use the χ^2 test for goodness of fit.

While running our simulation hopefully gave you a better idea of how the χ^2 statistic is created and distributed, a more accurate and efficient way of performing a χ^2 test for goodness of fit uses a table of chi-square critical values, a calculator or a computer. Table 10.1 is a table of chi-square critical values. For reference, this table can also be found as Table C at the end of the book.

Significance Level α

df	0.25	0.2	0.15	0.1	0.05	0.01	0.001
1	1.32	1.64	2.07	2.71	3.84	6.63	10.83
2	2.77	3.22	3.79	4.61	5.99	9.21	13.82
3	4.11	4.64	5.32	6.25	7.81	11.34	16.27
4	5.39	5.99	6.74	7.78	9.49	13.28	18.47
5	6.63	7.29	8.12	9.24	11.07	15.09	20.52
6	7.84	8.56	9.45	10.64	12.59	16.81	22.46
7	9.04	9.80	10.75	12.02	14.07	18.48	24.32
8	10.22	11.03	12.03	13.36	15.51	20.09	26.12
9	11.39	12.24	13.29	14.68	16.92	21.67	27.88

Table 10.1

This table tells you, given degrees of freedom, how large a χ^2 statistic must be to be significant at various levels.

Using the table, was our χ^2 statistic value for the die significant at the .05 level? _____ The .01 level? _____ The .001 level? _____

As we have found throughout our unit on inference, tables can be helpful but calculators and computers are more accurate and capable tools. The two major advantages of using a calculator or computer when performing a chi-square test for goodness of fit are that 1, we can enter any degrees of freedom and 2, the result will be an exact P-value. This means that if there is a 4.2% chance of getting a chi-square value of some size, the calculator will output .042. The table will merely tell us that the chi-square statistic value was large enough to be significant at the .05 level but not at the .01 level.

To perform a chi-square goodness of fit test on a TI-84 calculator (sadly, the TI-83 does not have the χ^2 GOF-Test as an option), we begin by entering out observed counts in L1 and our expected counts in L2. For our die data, that means that our screen should look like the following:

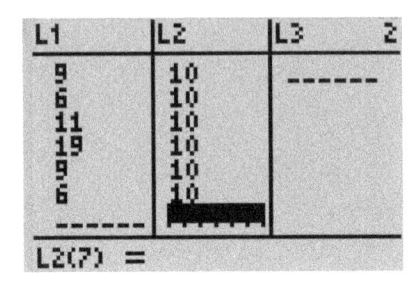

Next, press STAT, arrow over to TESTS, and scroll down to select D: χ^2 GOF-Test. Enter L1 for Observed, L2 for Expected, 5 for df and then select Draw.

What P-value did the calculator find for our die data? _____

Homework Assignment 10.4: Testing the Tootsie Pop Distribution

Sweets Company of America claims that Tootsie Pops are 20% chocolate, 20% cherry, 20% orange, 10% grape, 10% raspberry, 10% strawberry, 5% watermelon, and 5% blue raspberry. Of the 200 Tootsie Pops that Mr. Davidson gave out this year, there were 32, 41, 47, 29, 6, 18, 9, and 3 respectively for those flavors. Let's perform a χ^2 test for goodness of fit to see if we should believe Sweets' claimed flavor distribution.

State our null and alternative hypotheses:

H_0: SCA claim = CORRECT

H_a: SCA claim = FALSE

Find the expected counts for each flavor:

1. Chocolate: $\frac{(32-40)^2}{40} = 1.6$
2. Cherry: $\frac{(41-40)^2}{40} = .025$
3. Orange: $\frac{(47-40)^2}{40} = 1.225$
4. Grape: $\frac{(29-20)^2}{20} = 4.05$
5. Raspberry: $\frac{(6-20)^2}{20} = 9.8$
6. Strawberry: $\frac{(18-20)^2}{20} = 0.2$
7. Watermelon: $\frac{(9-10)^2}{10} = 0.1$
8. Blue Raspberry: $\frac{(3-10)^2}{10} = 4.9$

Calculate χ^2: $\boxed{21.9}$

\# of trials × probability of success of outcome

x	1	2	3	4	5	6	7	8
$p(x)$	.2	.2	.2	.1	.1	.1	.05	.05

$200(.2) =$

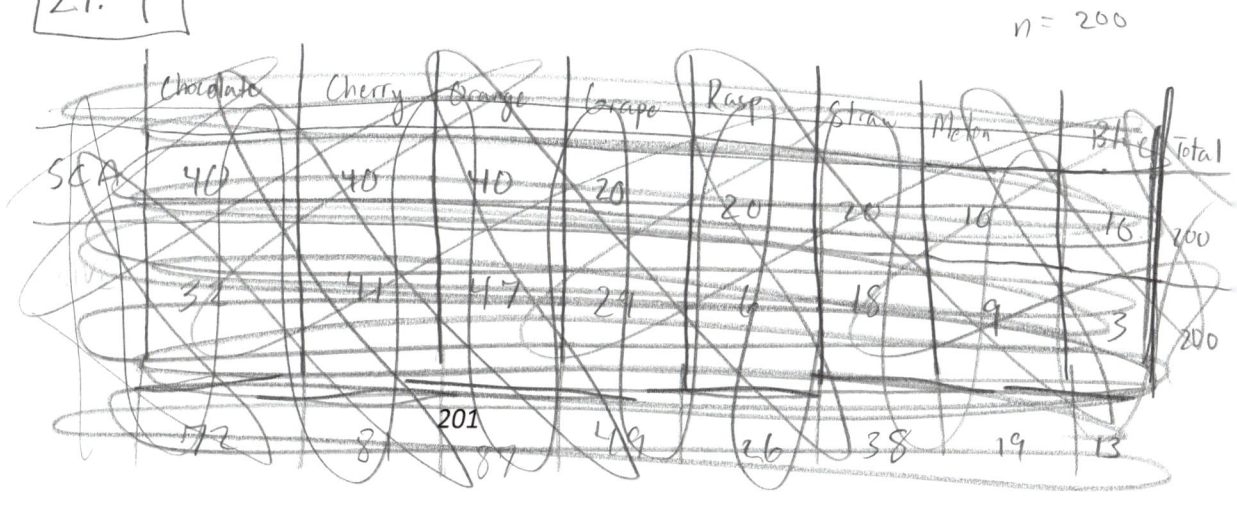

$n = 200$

Find the degrees of freedom:

Use Table 10.1 to determine the smallest α for which our χ^2 is significant:

Do you think we should believe Sweets' proposed flavor distribution? Why?
_____
_____
_____

Class Activity 10.6: The Chi-Square Test for Homogeneity and Independence

In Chapter 7, we used two-way tables to display the sample space of random phenomena when two events are being studied. Two way tables can also be used to show the relationship between two categorical variables measured within a population. For example, let's say that junior school students and parents are worried that athletic demands will be more substantial in the high school and they fear that it will be tough to manage time. They select students at random and put their collected data in the following two-way table.

		Athletic Hours/Week		
		< 3	3 – 10	> 10
School	HS	10	61	31
	JS	7	31	10

We were able to test the null hypothesis that the die was fair and that Sweets' flavor distribution was accurate using a chi-square test for goodness of fit. To test the null hypothesis that athletic demands in the junior school and high school are equivalent, we can do a chi-square test for a two-way table. There are two chi-square tests that are run on two way tables, the **chi-square test for homogeneity** and the **chi-square test for independence**. The two tests are identical except for their design. If data is collected by performing separate random samples for the two groups being studied, it is a chi-square test for homogeneity. If one random sample is taken from the entire population and both categorical variables are recorded, it is a chi-square test for independence.

If this two-way table were created by taking a random sample of 102 high school students and then another random sample of 48 junior school students and their athletic demands were recorded for both, which type of chi-square test would we be performing?

_____

If this two-way table were created by taking a random sample of 150 MBA students are recording whether they were in high school or junior school and their athletic demands, which type of chi-square test would we be performing? _____

The process of testing a null hypothesis given a two-way table is the same regardless of whether data was collected using one or two random samples. The null hypothesis (H_0) is that there is no association between the two categorical variables. The alternative hypothesis (H_a) is that there is. To find the expected counts assuming H_0 is true in any cell in a two-way table, you use the following formula:

$$expected\ count = \frac{row\ total\ \times\ column\ total}{table\ total}$$

The degrees of freedom when using a two-way table can be found with the following formula:

$$degrees\ of\ freedom = (r-1)(c-1)$$

where r = # of rows and c = # of columns.

The chi-square statistic formula remains the same and the same chi-square critical values table can be used to find the alpha level at which the chi-square statistic value is significant.

Let's perform a chi-square test on the two-way table above. First, let's state our null and alternative hypotheses:

H₀:

Hₐ:

Next, let's find the expected counts in all 6 cells. The first cell has been calculated for you. 11.56 = (102·17)/150.

		Athletic Hours/Week		
		< 3	3 – 10	> 10
School	HS	11.56	_____	_____
	JS	_____	_____	_____

Next, let's calculate the chi-square statistic value. As a reminder:

$$\chi^2 = \sum \frac{(observed - expected)^2}{expected}$$

$\chi^2 =$

Now, let's find the degrees of freedom.

Lastly, use the Chi-square critical values table (Table C) to find the smallest alpha for which our chi-square statistic value is significant.

Homework Assignment 10.5: Using a Calculator to Perform Two-Way Table Chi-Square Test

Let's find the P-value for our athletic commitment two-way table using our calculator. First, press 2nd, x^{-1} to select MATRIX, cursor over to EDIT and select 1: [A]. Specify a 2x3 matrix and enter our observed counts in the appropriate cells. When you are finished, your calculator should look like the following:

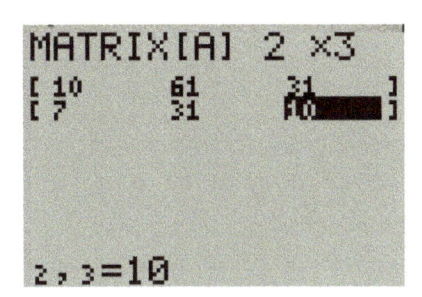

Now press STAT, move over to TESTS and choose C: χ^2-Test. Both "Calculate" and "Draw" will find the P-value for you. Choose one and record the P-value that the calculator finds for you. __0.392__

Class Activity 10.7: Inference about a Population Mean: Ages at a Little League Game

We discussed confidence intervals for proportions in Chapter 9. Now we will discuss confidence intervals about a population mean. That means that instead of confidence statements like "I am 90% confident that 55% of people support lower taxes with a 4% margin of error" we'll make statements like "I am 98% confident that the average weight of an MBA student is 178 pounds with a margin of error of 8 pounds."

In some cases, like the weight of MBA students, the population distribution might be approximately normal. Let's consider a case where the population distribution is not normal. Let's say we're trying to determine the average age at a Little League baseball game. The vast majority of people in attendance are players and their parents or guardians. We're going to use an online applet developed by David Lane at Rice University to see what happens when we draw samples from this baseball game population. Go to the following website:

www.onlinestatbook.com/stat_sim/sampling_dist/index.html

Select BEGIN. You should see a normal distribution near the top of the screen. We want to create a histogram that shows our baseball game ages distribution so you should click the "Normal" next to the graph and select "Custom" from the list of options. After the graph is cleared you can use your mouse to create the baseball game age distribution you would find at the game. You can see below the histogram that I created.

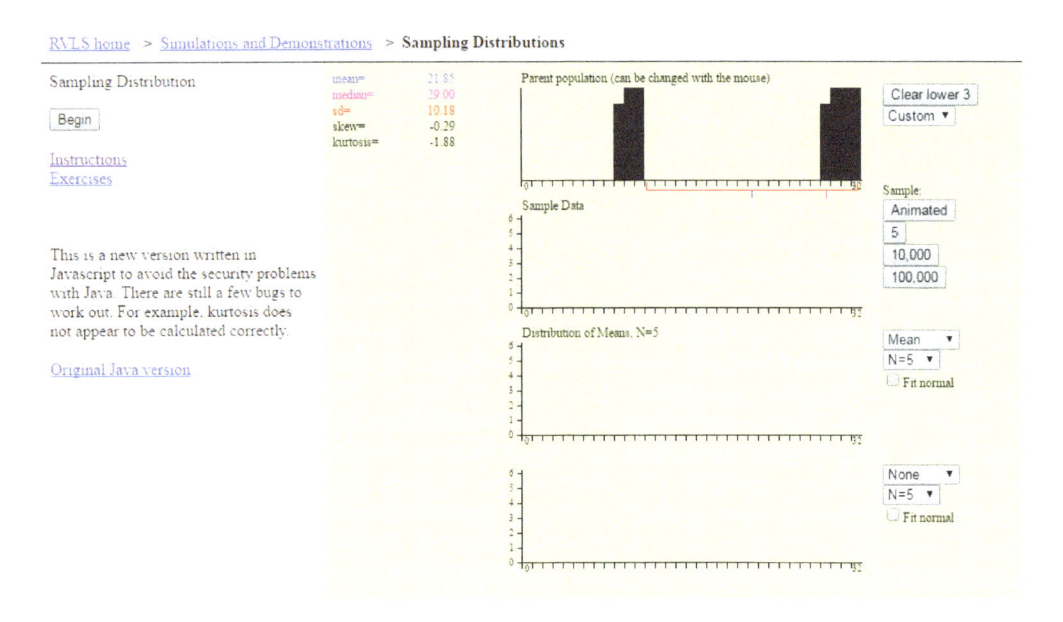

Next to the third graph make sure "Mean" and "N=2" are selected. This means that we will be taking a sample size of two and recording the sample mean with every repetition. Click the "Animated" button and the Applet will show two randomly selected individuals from the population being selected and then their average being placed in the third graph. Continue pushing "Animated" until you see a pattern. What do you see?

_____
_____

Now, let's make our sample size larger. Change "N=2" to "N=5" and instead of hitting "Animated," push the "10,000" button to simulate 10,000 repetitions. What do you see?

_____
_____

Finally, let's make our sample size "N=25" and push "10,000" again. This time, what do you see?

_____
_____

What we have just witnessed is the **central limit theorem** in action. The central limit theorem says that with a large enough sample size, even populations that are not normal will have sampling distributions of the sample means that are approximately normal.

Homework Assignment 10.6: The Central Limit Theorem and Standard Deviation of a Sampling Distribution

We looked at ages of people at a Little League baseball game. Think of another distribution that is not normal. What distribution do you want to explore for this activity?

ages of people at an MLB game

Draw this distribution on the website. Look at the distribution of sample means as sample size increases from 2 to 25 to confirm that this mean distribution also becomes approximately normal as the sample size increases.

Select "N=5" and push "100,000" and record the mean and standard deviation of this distribution. These values are provided to the left of the graph. Mean = __26__.
Standard Deviation = __2.24__.

Select "N=20" and push "100,000" and record the mean and standard deviation of this distribution. These values are provided to the left of the graph. Mean = __16__.
Standard Deviation = __1.12__.

Given these two results, what do you think is the relationship between the mean of the population (μ) and the mean of the distribution of the sample means ($\mu_{\bar{x}}$)?

they remain constant

Given these two results, what do you think is the relationship between the standard deviation of the population (σ) and the standard deviation of sampling distribution ($\sigma_{\bar{x}}$)?

σ twice the sampling distribution

Class Activity 10.8: Making a Confidence Interval for a Population Mean

When you looked at the mean and sampling distribution of $\bar{x}$ you likely noticed that the mean stayed the same and that the standard deviation halved as you changed from a sample size of 5 to 20. This is because of the following relationship between the mean and standard deviation of a population and the mean and standard deviation of a sample drawn from that population:

$$\mu_{\bar{x}} = \mu$$

and

$$\sigma_{\bar{x}} = \frac{\sigma}{\sqrt{n}}$$

With those two facts combined with our central limit theorem which tells us that the sampling distribution of sample means from any population is approximately normal if the sample size is large enough (the typical requirement is n ≥ 30), we can make a confidence interval for a population mean. Just like our confidence interval for a proportion, our basic form is estimate ± margin of error with a certain level of confidence, but for means instead of proportions that means our interval is

$$\bar{x} \pm z* \frac{\sigma}{\sqrt{n}}$$

But there's a problem. Just like when we wanted to make a confidence interval for proportions and the correct formula involved p (an unknown value), this formula also uses a value that we would normally not know when taking a sample. Which value would be unknown when taking a sample from a population? _____

As a result, we need to use our sample standard deviation (s) in place of the population standard deviation (σ) to make an approximate confidence interval. This means that if an SRS of size n is taken from a large population, when n is large (≥ 30) then an approximate level C confidence interval for μ is

$$\bar{x} \pm z* \frac{s}{\sqrt{n}}$$

where z* is the critical value for confidence level C that can be found in Table B at the end of the book.

Let's say that 31 people are randomly selected at the Little League baseball game. The mean age in the sample is 20.97 years and the standard deviation of the ages in the sample is 6.41 years. Use our formula and Table B to find a 90% confidence interval for μ.

5.9161

1.8593

4.7970

Homework Assignment 10.7: Practicing Mean Confidence Intervals

Let's say that 51 students are randomly selected from the MBA student body at the end of a school year and are asked how many hours of service they did that year. The mean number of hours in the sample is 35 hours and the standard deviation of the hours in the sample is 11 hours. Use our formula and Table B to find a 99% confidence interval for µ.

$$\bar{x} \pm z^* \frac{\sigma}{\sqrt{n}}$$

$$35 + 2.58 \left(\frac{11}{\sqrt{51}}\right) = 57.884$$

to

$$35 - 2.58 \left(\frac{11}{\sqrt{51}}\right) = 49.937$$

Class Activity 10.9: The T Interval: Better for Small Samples and Larger

William Sealy Gosset, an English statistician, was working for the Guinness brewery in Dublin in the late 19th and early 20th century when he noticed a problem with the confidence interval formula we introduced in Class Activity 10.8. While distributions are approximately normal for large sample sizes, when sample sizes are small, replacing σ with s and assuming the distribution is normal can lead to inaccurate results. This led Gosset to develop the **t distributions**. There is a different t distribution for every sample size n and the t distributions more accurately capture the distribution of sample means than does a normal distribution. While t distributions are really necessary when dealing with small sample sizes, they are also more accurate even when the sample size is large, i.e. when n ≥ 30. This means that confidence interval for a population mean can be found using t distributions and the following formula:

$$\bar{x} \pm t^* \frac{s}{\sqrt{n}}$$

where t* is the critical value for the t distribution with n – 1 degrees of freedom.

Let's find a confidence interval for our Little League baseball game again, but this time let's use the t distribution for more accurate results. As a reminder, 31 people were randomly selected at the Little League baseball game. The mean age in the sample was 20.97 years and the standard deviation of the ages in the sample was 6.41 years. Use our new formula and Table D to find a 90% confidence interval for μ.

Homework Assignment 10.8: A Better Confidence Interval With t*

Let's revisit our service at MBA question. Remember that 51 students were randomly selected from the MBA student body at the end of a school year and are asked how many hours of service they did that year. The mean number of hours in the sample was 35 hours and the standard deviation of the hours in the sample was 11 hours. Use our t distribution formula and Table D to find a 99% confidence interval for µ.

Class Activity 10.10: The T Distribution on the Calculator

You likely realized that the sample sizes of 31 and 51 in the two examples were intentionally selected so that critical values for the degrees of freedom, 30 and 50 respectively, would be available in Table D. If the degrees of freedom are not available in your table or you want more accurate results, it is good to use a calculator or computer. Let's calculate the same confidence interval using the t distribution for 50 degrees of freedom in our Service Club problem.

Press STAT, move over to TESTS, and select 8:TInterval. You can enter data into a list and find the confidence interval given that data, or if you know the mean, standard deviation and sample size you can select Stats and enter information there. Since the latter is our case, select Stats and enter in 35, 11, 51 and .99 for the sample mean, sample standard deviation, same size and confidence level respectively. When you are done, you screen should look like the following.

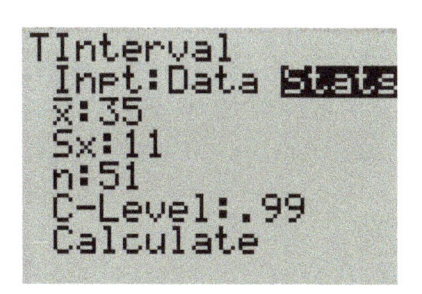

```
TInterval
  Inpt:Data Stats
  x:35
  Sx:11
  n:51
  C-Level:.99
  Calculate
```

Select Calculate and press ENTER. What 99% confidence interval did the calculator find?
_____

Was this the same result that we got using Table D in Homework Assignment 10.8?
_____

Homework Assignment 10.9: Chapter 10 Review Exercises

Let's say that Coach McMahan claims to be a 91% free throw shooter. To test this theory, a Statistics class asks him to shoot some free throws. Coach McMahan makes 90% of the free throws he attempts. It is possible that this result would be enough to reject the null hypothesis that Coach McMahan is a 91% free throw shooter at the α=.01 level. What would have to be the case about the number of free throws Coach McMahan shot for this to be true? _____

If this condition were met and the null hypothesis were rejected at the α=.01 level, it would be a statistically significant result. Do you think it would also be a practically significant result? Why? _____
_____
_____
_____
_____

A student is repeatedly selected for drug testing at MBA. He argues that he has been selected for testing so frequently that it is improbable that selection is truly random. He does a significance test and finds a P-value just under .01. Since this result is significant at the α=.01 level, he feels that this is sufficient evidence to reject the null hypothesis that drug testing is done at random at MBA. Does the student have a good case? Why or why not? _____
_____
_____
_____
_____

There are a number of inference tests that we did not discuss in this textbook but they are all based on the same idea. Assume some null hypothesis is true and determine how often a sample result like the one collected would occur given that null hypothesis assumption. The less likely it is, the more evidence we have to reject the null hypothesis. For example, consider the case of Tim Donaghy, the NBA referee who in July 2007 was accused of fixing NBA games. The claim was that he was calling more fouls in fixed games in the hopes of increasing the score and helping those gamblers who bet that the total points scored would be more than the total points line. When games are fixed and many betters take the over, the line often shifts as a result. Statisticians compared the games Tim Donaghy refereed where the line shifted up two or more points to the games he refereed where the line did not shift up two or more points. They found that in the 11 games Donaghy officiated in which the total line shifted up two or more points, 180.27 more free throws than expected were attempted. Statisticians assumed that Tim Donaghy was calling the games fairly and then calculated the likelihood that this kind of deviation would occur in these games where the line changed dramatically. They determined that there was approximately a .005 probability that the change in free throws was merely the result of chance. This small P-value was small enough to convince most that Tim Donaghy was fixing games.[43] While we did not cover two sample inference tests, what the statisticians were trying to do and why the result was strong evidence against Donaghy's claims of innocence should both be clear now that you've had this introduction to inference. What other data could statisticians study to try to determine if Donaghy was fixing games?

_____
_____
_____
_____

Index

A

alternative hypothesis, 185, 187, 188, 191, 192, 203

B

bar graph, 6, 46, 48, 74
bias, 7, 11, 12, 15, 16, 17, 27, 28, 29, 32, 34, 124
bimodal, 55, 56, 57
binary, 161, 195
binomial setting, 161, 162, 163, 166, 189
binomial theorem, 130, 159, 160, 161, 163
block design, 38
boxplot, 60, 61, 62, 63

C

categorical, 2, 4, 47, 48, 195, 203
census, 5, 11, 28
central limit theorem, 206, 207
chi-square (χ^2) test for goodness of fit, 195
chi-square test for homogeneity, 203
chi-square test for independence, 203
Clinical trials, 33
cluster sampling, 26
coefficient of determination, 115, 117, 124
combinations, 157, 158, 159
common response, 121, 122
confidence statement, 22, 25, 26, 183
confidential, 42
confounded, 31, 39, 121, 123
continuous, 2, 3, 4, 75, 78, 82
control group, 7, 8, 32, 33, 35, 37, 44
convenience sampling, 11
correlation, 105, 106, 107, 108, 109, 110, 111, 112, 114, 115, 121, 122, 124, 125
critical values, 180, 181, 199, 204, 211, 219, 220

D

degrees of freedom, 198, 199, 202, 204, 210, 211
density curve, 78, 80, 82, 88, 90, 198
direction, 50, 79, 103, 110, 187
discrete, 2, 3, 4, 75, 82, 83
distribution, 45, 46, 47, 55, 56, 57, 59, 60, 63, 75, 76, 77, 78, 80, 81, 82, 83, 84, 85, 87, 88, 89, 90, 91, 92, 93, 94, 95, 96, 99, 151, 152, 153, 164, 165, 166, 173, 175, 176, 178, 179, 181, 186, 187, 188, 189, 191, 195, 197, 198, 201, 202, 203, 205, 207, 210, 211, 220
dotplots, 47
double-blind, 7, 8, 34, 35
dropouts, 8, 34

E

empirical rule, 88
event, 127, 128, 133, 140, 141, 160
experiments, 1, 8, 33, 34, 37, 38, 39, 41, 44, 121
Experiments, 1, 8, 31
explanatory variable, 31, 101, 104, 111, 117, 124

F

factorial, 155
false negative, 143
false positive, 143
first quartile, 59
five-number summary, 59, 60, 61, 62, 63
form, 26, 42, 82, 85, 103, 104, 151, 153, 154, 208

G

gambler's fallacy, 127

H

histograms, 47, 58, 78, 80

I

independent, 31, 101, 128, 130, 133, 138, 140, 145, 152, 159, 161, 166, 191
individuals, 2, 4, 11, 12, 17, 18, 26, 28, 31, 40, 55, 75, 77, 97, 101, 105, 106, 117, 124, 156, 157, 160, 177, 187, 206
institutional review board, 42, 43
interquartile range, 60

L

Law of Large Numbers, 127, 129, 130, 131, 147, 169
least-squares regression, 110, 111, 125
level of confidence, 22, 30, 180, 208
line graph, 71, 74
lurking variable, 31, 38, 41, 121, 122

215

M

margin of error, 22, 23, 24, 25, 28, 29, 30, 175, 176, 205, 208
matched pairs design, 37, 38
matching, 41
maximum, 59, 90, 105
mean, 8, 28, 35, 42, 63, 64, 67, 76, 77, 79, 80, 82, 88, 90, 91, 93, 94, 95, 97, 99, 113, 114, 117, 122, 125, 166, 173, 175, 176, 179, 180, 181, 186, 187, 188, 191, 193, 205, 206, 207, 208, 209, 210, 211
median, 39, 55, 56, 57, 59, 60, 63, 78, 80, 82
minimum, 59
mode, 55, 63
multimodal, 55
multiplication principle, 154, 156
mutually exclusive, 133

N

negative association, 103
nonadherers, 34
nonresponse, 28, 34
nonsampling errors, 28, 34
normal distribution, 82, 83, 88, 91, 92, 99, 176, 210
null hypothesis, 185, 186, 187, 188, 191, 192, 193, 195, 203, 212

O

observational studies, 1, 41
one-sided alternative hypothesis, 187
outlier, 57, 60, 62, 102, 104, 114, 122
outliers, 56, 60, 61, 62, 63, 103, 104, 114

P

parameter, 11, 12, 15, 17, 18, 21, 22, 25, 28, 34, 175, 176, 177, 178
Pascal's Triangle, 152, 153, 154
percentile, 59, 75, 77, 88, 90, 91, 94, 95
permutations, 156, 158
pictoral graph, 74
pie chart, 6, 46
placebo effect, 32, 33, 44
population, 5, 7, 11, 12, 14, 15, 16, 17, 18, 21, 22, 23, 24, 25, 26, 27, 28, 29, 34, 35, 41, 46, 70, 71, 72, 77, 89, 90, 91, 124, 166, 173, 175, 177, 178, 180, 195, 203, 205, 206, 207, 208, 210
positive association, 103

power of the hypothesis test, 192, 193
practical significance, 35
probability, 27, 83, 89, 91, 95, 99, 127, 128, 129, 130, 131, 132, 133, 137, 138, 139, 140, 141, 143, 144, 147, 149, 150, 151, 152, 153, 154, 155, 158, 159, 160, 161, 162, 163, 164, 165, 166, 167, 168, 169, 170, 171, 172, 173, 181, 185, 186, 187, 188, 189, 190, 192, 193, 213, 220
probability samples, 27
processing errors, 28
pseudo-random numbers, 16
P-value, 185, 186, 187, 188, 189, 190, 191, 192, 199, 200, 205, 212, 213

Q

quantitative, 2, 3, 4, 47, 48, 71, 101, 124
quick method for margin of error, 22, 25, 30

R

random, 7, 8, 9, 12, 13, 15, 16, 17, 22, 26, 27, 28, 30, 32, 33, 37, 38, 39, 83, 95, 96, 124, 127, 129, 130, 131, 132, 133, 137, 151, 152, 153, 155, 159, 162, 164, 166, 168, 179, 187, 191, 203, 212
random number generator, 12, 13, 37
random sampling error, 28, 30
randomized comparative experiment, 33, 37, 39, 40, 44
range, 2, 22, 23, 48, 56, 57, 58, 59, 60, 66, 84, 91, 93, 101, 102, 113, 124
refusals, 34
regression, 107, 110, 111, 112, 113, 114, 115, 117, 118, 123, 124, 125
Regression, 110, 111, 114, 120, 122
residual, 117, 118, 119, 120, 125
response errors, 28
response variable, 31, 101, 104, 111, 117, 121, 124

S

sample, 1, 5, 7, 9, 11, 12, 15, 16, 17, 18, 19, 21, 22, 23, 25, 26, 27, 28, 29, 33, 34, 63, 64, 68, 69, 70, 71, 77, 84, 85, 91, 115, 133, 134, 140, 164, 166, 175, 177, 178, 179, 185, 187, 188, 189, 193, 195, 198, 203, 206, 207, 208, 209, 210, 211, 212
sampling distribution, 166, 175, 186, 207
sampling errors, 28
sampling frame, 25, 28, 29
sampling variability, 7, 17, 28

scatterplot, 101, 102, 103, 104, 105, 110, 112, 113, 114, 117, 118, 124, 125
seasonal variation, 74
significance level, 187
significance test, 185, 186, 195, 212
simple random sample, 12, 26
Simpson's Paradox, 142
simulation, 96, 98, 99, 130, 131, 196, 197, 199
skewed left, 56, 57, 79, 82, 197
skewed right, 56, 57, 197
standard deviation, 39, 63, 64, 65, 68, 69, 70, 71, 76, 77, 82, 88, 90, 91, 93, 94, 95, 97, 99, 117, 166, 173, 175, 176, 177, 178, 179, 180, 186, 187, 207, 208, 209, 210, 211
standard normal distribution, 91
statistic, 11, 15, 25, 26, 28, 34, 165, 166, 173, 175, 177, 178, 186, 188, 195, 196, 197, 199, 204
statistically significant, 7, 8, 35, 36, 187, 212
Statistics, 1, 31, 39, 73, 74, 150, 154, 156, 212
stemplots, 47
stratified random sampling, 26, 38
strength, 103, 104, 105, 110, 117
subjects, 1, 7, 8, 9, 31, 32, 33, 34, 37, 38, 39, 41, 42, 43, 44
surveys, 5, 29, 34
symmetric, 56, 57, 63, 79, 82, 85, 168, 197

T

table of random digits, 12, 13, 37
third quartile, 59
treatment, 1, 31, 32, 33, 34, 35, 37, 38, 39, 41, 43, 44
tree diagram, 137, 139, 148, 149
trend, 72, 74, 110, 122
two-sided alternative hypothesis, 187
two-way table, 134, 135, 136, 143, 203, 204, 205
Type I error, 192, 193
Type II error, 192, 193

U

undercoverage, 28
uniform distribution, 83
unimodal, 55, 56, 57

V

variables, 2, 3, 4, 31, 32, 33, 37, 38, 39, 40, 41, 47, 75, 82, 99, 101, 102, 103, 104, 105, 106, 110, 114, 117, 121, 124, 125, 151, 195, 203
variance, 64, 68, 71, 99
Venn diagram, 135
voluntary informed consent, 42, 44
voluntary response, 12, 15, 27

Z

z-score, 76, 77, 91, 181, 188, 190, 191
Z-test, 186, 191, 192, 195

Tables
Table A: Standard Normal Probabilities

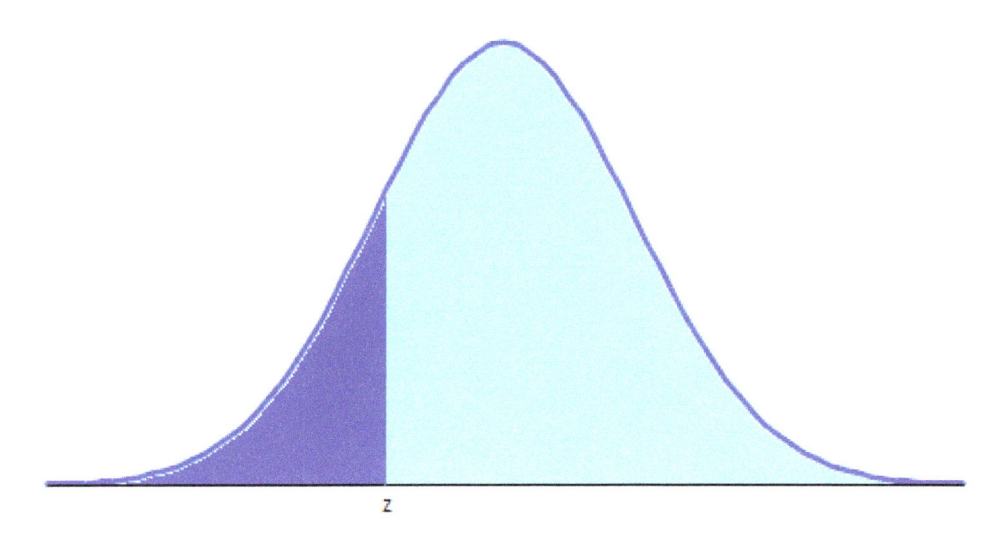

Table entry is the area under the standard normal curve to the left of the corresponding z value

z		z		z		z	
-3.0	0.0013	-1.5	0.0668	0.0	0.5000	1.5	0.9332
-2.9	0.0019	-1.4	0.0808	0.1	0.5398	1.6	0.9452
-2.8	0.0026	-1.3	0.0968	0.2	0.5793	1.7	0.9554
-2.7	0.0035	-1.2	0.1151	0.3	0.6179	1.8	0.9641
-2.6	0.0047	-1.1	0.1357	0.4	0.6554	1.9	0.9713
-2.5	0.0062	-1.0	0.1587	0.5	0.6915	2.0	0.9772
-2.4	0.0082	-0.9	0.1841	0.6	0.7257	2.1	0.9821
-2.3	0.0107	-0.8	0.2119	0.7	0.7580	2.2	0.9861
-2.2	0.0139	-0.7	0.2420	0.8	0.7881	2.3	0.9893
-2.1	0.0179	-0.6	0.2743	0.9	0.8159	2.4	0.9918
-2.0	0.0228	-0.5	0.3085	1.0	0.8413	2.5	0.9938
-1.9	0.0287	-0.4	0.3446	1.1	0.8643	2.6	0.9953
-1.8	0.0359	-0.3	0.3821	1.2	0.8849	2.7	0.9965
-1.7	0.0446	-0.2	0.4207	1.3	0.9032	2.8	0.9974
-1.6	0.0548	-0.1	0.4602	1.4	0.9192	2.9	0.9981
						3.0	0.9987

Table B: Critical Values (z*) of the Normal Distribution

Confidence Level C	Critical Value z*	Confidence Level C	Critical Value z*
50%	0.67	90%	1.64
60%	0.84	95%	1.96
70%	1.04	99%	2.58
80%	1.28	99.9%	3.29

Table C: Chi-square critical values

	\multicolumn{7}{c}{Significance Level α}
df	0.25	0.2	0.15	0.1	0.05	0.01	0.001
1	1.32	1.64	2.07	2.71	3.84	6.63	10.83
2	2.77	3.22	3.79	4.61	5.99	9.21	13.82
3	4.11	4.64	5.32	6.25	7.81	11.34	16.27
4	5.39	5.99	6.74	7.78	9.49	13.28	18.47
5	6.63	7.29	8.12	9.24	11.07	15.09	20.52
6	7.84	8.56	9.45	10.64	12.59	16.81	22.46
7	9.04	9.80	10.75	12.02	14.07	18.48	24.32
8	10.22	11.03	12.03	13.36	15.51	20.09	26.12
9	11.39	12.24	13.29	14.68	16.92	21.67	27.88

Table D: t distribution critical values

				Upper tail probability p					
df	0.25	0.1	0.05	0.025	0.01	0.005	0.0025	0.001	0.0005
1	1.000	3.078	6.314	12.71	31.82	63.66	127.3	318.3	636.6
2	0.816	1.886	2.920	4.303	6.965	9.925	14.09	22.33	31.60
3	0.765	1.638	2.353	3.182	4.541	5.841	7.453	10.21	12.92
4	0.741	1.533	2.132	2.776	3.747	4.604	5.598	7.173	8.610
5	0.727	1.476	2.015	2.571	3.365	4.032	4.773	5.893	6.869
6	0.718	1.440	1.943	2.447	3.143	3.707	4.317	5.208	5.959
7	0.711	1.415	1.895	2.365	2.998	3.499	4.029	4.785	5.408
8	0.706	1.397	1.860	2.306	2.896	3.355	3.833	4.501	5.041
9	0.703	1.383	1.833	2.262	2.821	3.250	3.690	4.297	4.781
10	0.700	1.372	1.812	2.228	2.764	3.169	3.581	4.144	4.587
11	0.697	1.363	1.796	2.201	2.718	3.106	3.497	4.025	4.437
12	0.695	1.356	1.782	2.179	2.681	3.055	3.428	3.930	4.318
13	0.694	1.350	1.771	2.160	2.650	3.012	3.372	3.852	4.221
14	0.692	1.345	1.761	2.145	2.624	2.977	3.326	3.787	4.140
15	0.691	1.341	1.753	2.131	2.602	2.947	3.286	3.733	4.073
16	0.690	1.337	1.746	2.120	2.583	2.921	3.252	3.686	4.015
17	0.689	1.333	1.740	2.110	2.567	2.898	3.222	3.646	3.965
18	0.688	1.330	1.734	2.101	2.552	2.878	3.197	3.610	3.922
19	0.688	1.328	1.729	2.093	2.539	2.861	3.174	3.579	3.883
20	0.687	1.325	1.725	2.086	2.528	2.845	3.153	3.552	3.850
21	0.686	1.323	1.721	2.080	2.518	2.831	3.135	3.527	3.819
22	0.686	1.321	1.717	2.074	2.508	2.819	3.119	3.505	3.792
23	0.685	1.319	1.714	2.069	2.500	2.807	3.104	3.485	3.768
24	0.685	1.318	1.711	2.064	2.492	2.797	3.091	3.467	3.745
25	0.684	1.316	1.708	2.060	2.485	2.787	3.078	3.450	3.725
26	0.684	1.315	1.706	2.056	2.479	2.779	3.067	3.435	3.707
27	0.684	1.314	1.703	2.052	2.473	2.771	3.057	3.421	3.690
28	0.683	1.313	1.701	2.048	2.467	2.763	3.047	3.408	3.674
29	0.683	1.311	1.699	2.045	2.462	2.756	3.038	3.396	3.659
30	0.683	1.310	1.697	2.042	2.457	2.750	3.030	3.385	3.646
50	0.679	1.299	1.676	2.009	2.403	2.678	2.937	3.261	3.496
100	0.677	1.290	1.660	1.984	2.364	2.626	2.871	3.174	3.390
1000	0.675	1.282	1.646	1.962	2.330	2.581	2.813	3.098	3.300
z*	0.674	1.282	1.645	1.960	2.326	2.576	2.807	3.091	3.291
Confidence Level C	50%	80%	90%	95%	98%	99%	99.5%	99.8%	99.9%

[1] National Center for Education Statistics (December 2012). "Table 5 Number of educational institutions, by level and control of institution: Selected years, 1980-81 through 2010-11". U.S. Department of Education. Retrieved 14 January 2015.

[2] Number of Majors and Athletic Division from The Princeton Review (2014) Academics & Majors and Campus Life/Facilities http://www.princetonreview.com/college-education.aspx Retrieved 14 January 2015 and Distance from Google Maps https://www.google.com/maps

[3] Found at http://www.vucommodores.com/sports/m-baskbl/mtt/vand-m-baskbl-mtt.html

[4] Anderson TW, Reid DBW, Beaton GH. Vitamin C and the common cold: a double-blind trial. Canadian Medical Association Journal 1972;107(6):503-508. http://www.ncbi.nlm.nih.gov/pmc/articles/PMC1940935/?page=3

[5] For more information on the Ann Landers survey, see the following article written by Western University http://www.stats.uwo.ca/faculty/bellhouse/stat353annlanders.pdf

[6] See more 2010 Census data about Nashville at the US Government's Census Bureau's database at http://quickfacts.census.gov/qfd/states/47/4752006.html

[7] Cynthia Crossen, "Fiasco in 1936 Survey Brought 'Science' to Election Polling," Wall Street Journal, October 2, 2006.

[8] University of Pennsylvania's Math Department, "Case Study I: The 1936 Literary Digest Poll" http://www.math.upenn.edu/~deturck/m170/wk4/lecture/case1.html

[9] Gubernatorial polling data can be found at the poll aggregator Real Clear Politics, http://www.realclearpolitics.com/epolls/2014/governor/tn/tennessee_governor_haslam_vs_brown-5184.html#polls

[10] For the full report from Rasmussen, see the polling agency's write-up at http://www.rasmussenreports.com/public_content/politics/elections/election_2014/tennessee/election_2014_tennessee_governor

[11] Diane Whitmore Schanzenbach, "What Have Researchers Leaned from Project STAR?"Brookings Papers on Education Policy, August 2006

[12] Charles J. Wheelan, "Naked statistics: stripping the dread from the data." First Edition. P. 233-235.

[13] Stacy Berg Dale and Alan Krueger, "Estimating the Payoff to Attending a More Selective College: An Application of Selection on Observables and Unobservable," Quarterly Journal of Economics 117, no. 4 (November 2002): 1491-527.

[14] For more information about the Tuskeegee study, visit www.cdc.gov/tuskegee/.

[15] To read more about the Milgram Experiment, visit www.stanleymilgram.com

[16] More information about the Stanford Prison Experiment Is available at http://www.prisonexp.org/.

[17] For more information and data from recent surveys, visit www.bls.gov/cps/.

[18] From Open Source Shakespeare. For more information on Shakespeare's plays, visit http://www.opensourceshakespeare.org/.

[19] Salaries from Hoops Hype. http://hoopshype.com/salaries/memphis_grizzlies/

[20] Source: ESPN. http://espn.go.com/mlb/team/stats/batting/_/name/atl/year/2014/cat/atBats/atlanta-braves

[21] Census data tabulated at https://www.census.gov/population/www/censusdata/cencounts/files/tn190090.txt

[22] Box Office Mojo (http://www.boxofficemojo.com/alltime/domestic.htm), July 7, 2015.

[23] Box Office Mojo (http://www.boxofficemojo.com/alltime/adjusted.htm), July 7, 2015.

[24] Darrell Huff, (1954) How to Lie with Statistics (illust. I. Geis), Norton, New York, pp. 71-72.

[25] Wolfers, Justin. "Point Shaving in NCAA Basketball." American Economic Review 96, no. 2 (2006): 279-283.

[26] Bernhardt, D. and Heston, S. (2010), Point Shaving in College Basketball: A Cautionary Tale for Forensic Economics. Economic Inquiry, 48: 14–25. doi: 10.1111/j.1465-7295.2009.00253.x

[27] For the full article "So You Want to Play Pro Basketball" go to the following link. http://www.nytimes.com/interactive/2013/11/03/sunday-review/so-you-want-to-play-pro-basketball.html?_r=1&

[28] For the rest of the 2014 Percentile Ranks, check out https://secure-media.collegeboard.org/digitalServices/pdf/sat/sat-percentile-ranks-crit-reading-math-writing-2014.pdf

[29] Mean approximated from PGA Tour data http://www.pgatour.com/players/player.27649.brandt-snedeker.html/statistics. Standard deviation simply chosen to make arithmetic easy for first problem using Table A.

[30] IQSP 2014 Conference Statistics. http://www.iqsp.net/confStatsLeaders.php?year=2014 November 14, 2015.

[31] HTTP://XKCD.COM/552/

[32] Mlodinow, Leonard. The Drunkard's Walk: How Randomness Rules Our Lives. New York: Vintage Books, 2008. Print, p. 175.

[33] Gilovich, T., R. Vallone, and A. Tversky. "The Hot Hand in Basketball: On the Misperception of Random Sequences." Cognitive Psychology 17 (1985): 295-314.

[34] This example is adapted from one in Charles Wheelan's "Naked Statistics: stripping the dread from the data." First Edition. P. 82-84.

[35] Charles Wheelan "Naked Statistics: stripping the dread from the data." First Edition. P. 80-82.

[36] https://www.tn.gov/assets/entities/revenue/attachments/f1314001Fill-in.pdf January 5, 2016.

[37] http://www.blazepizza.com/menu/nutrition-calculator/ January 5, 2016.

[38] This problem is adapted from one presented by Arthur Benjamin at the 2015 MathFest Conference in DC during the minicourse "The Mathematics of Games and Gambling"

[39] California Lottery. http://www.calottery.com/play/draw-games/powerball/powerball-changes January 10, 2016

[40] Jordan Ellenberg "How Not to be Wrong." First Edition. P. 133.

[41] Wolfram Research, Inc., Mathematica, Version 9.0, Champaign, IL (2012) used to create diagram.
[42] HTTP://XKCD.COM/882/
[43] Winston, Wayne L. "Mathletics: how gamblers, managers, and sports enthusiasts use mathematics in baseball, basketball and football." P. 244-247.